Agrarian Distress in India

ABOUT THE EDITORS

B.C. Barah, Dr. Barah is principal scientist (Economics) at National Centre for Agricultural Economics and Policy Research (NCAP). Experience of working in various institutions of repute including the Assam Agricultural University, International Crops Research Institute for semi arid & topics (ICRISAT), Central University of Hyderabad, NABARD, Cornell University, Duke University, (USA), IRRI, (Manila), Asian Productivity Organisation (Tokyo), NIPFP, NAIP and NFSM is asset to him. Dr. Barah has a high class Ph.D. degree in the applied econometrics and environment economics. Dr. Barah published more than eighty research papers in local, national and international journals, twenty technical reports/bulletins, six books, policy brief and policy papers. He participated widely as member of several policy forums, high powered committees both at the government and the CSO level. He is awarded the prestigious NABARD Chair Professorship. He supervised number of research students leading to M.Phil and Ph.D degree.

Smita Sirohi, Dr. Smita Sirohi is Principal Scientist, in the discipline of Agricultural Economics, at National Dairy Research Institute (NDRI), Karnal (Haryana). She has been actively engaged in research and teaching in the area of dairy economics. She had a brief period of serving at National Centre for Agricultural Economics and Policy Research (NCAP), New Delhi. She has guided a number of Masters and Ph.D. students, authored many research papers, book chapters, popular articles, review papers and has also written two books.

Agrarian Distress in India

Problems and Remedies

Edited by
B.C. Barah
and
Smita Sirohi

CONCEPT PUBLISHING COMPANY PVT. LTD.,
NEW DELHI-110059

ISBN-13: 978-81-8069-765-4

First Published 2011

Published and Printed by

Concept
Since 1974

Concept Publishing Company Pvt. Ltd.
Regd. Office:
A/15-16, Commercial Block, Mohan Garden,
New Delhi-110059 (INDIA)
Phones : 25351460, 25351794, *Fax* : 091-11-25357109
Email : publishing@conceptpub.com,
Website: www.conceptpub.com

Editorial Office:
H-13, Bali Nagar, New Delhi-110 015, India.

Cataloging in Publication Data--*Courtesy:* D.K. Agencies (P) Ltd. <docinfo@dkagencies.com>

Agrarian distress in India : problems and remedies / edited by B.C. Barah and Smita Sirohi.

p. cm.
Includes bibliographical references and index.
ISBN 9788180697654

1. Agriculture--Economic aspects--India--Congresses. 2. Agriculture and state--India--Congresses. 3. Farmers--India--Psychology--Congresses. I. Barah, B. C. II. Sirohi, Smita.

DDC 338.10954 22

Preface and Acknowledgements

Agriculture in India has witnessed tremendous progress in the green revolution period. It produced more than 233 million tons of food grains in 2008-09, that was bare 50 million tons in 1950s, it provides livelihood to more than 600 million rural population and the sector employs 120 million farm families. Having achieved the comfort and complacency of quantum jump in production propelled by ostracized green revolution, the sector has pass through several phases of intermittent growth with the terminal agrarian distress stage in recent period. During the past five decades, the food grain production increased phenomenally from 50.5 million tonnes in 1950s to 233 million tonnes in 2008-09, similarly rice from 20.6 million tonnes to 99.1 million tones, and wheat from 6.5 million tonnes to 80.6 million tonnes. Yet the performance of Indian agriculture has, not been consistent over the various distinct phases. In the growth trajectory, the production and productivity pathways experience unevenness. During the 1980s, the growth climbed the highest peak in food production at over 3.0 per cent per annum, but the same has been tumbled down during the 1990s and thereafter. The decadal growth rate declined to less than 2 per cent per annum during the 1990s, and the down trend continues till date. The stagnation in productivity, yield gap growing and the declining total factor productivity (TFP) indicate significant reduction in the impact of agricultural technology. At the same time, the cost of production escalates in violent manner caused by unabated soaring price of various inputs, as the farmer is a silent taker and not the maker. These resulted in extreme inter-regional disparity and variability in performances. The overall impact of such unfavourable situation is the erosion of farm income and insecurity, particularly as the off-farm opportunities are shrinking within the sector. At the same time, the farmers are generally trapped under the garb of indebtedness, which is the alleged cause of farmers' suicide in many parts of the country. Consequently, in order to escape the burnt of poverty and hunger, the flight of the population towards the urban centres is observed to be a common phenomenon in the rural India. The cause of such out-migration does not require lengthy debate as is seen in the back of the envelop calculation that the annual net farm income of an average farmer, even in a prosperous region like Punjab, is the tenth of the monthly salary of the fourth grade employee in the central sector.

Thus the situation calls for thorough analysis of the cause and effect relationship that is necessary to derive policy perspective of a national scale for income enhancement and livelihood in the rural areas. In view of the complexity, the situation requires proper understanding and insights. Realizing the gravity of the problems, the National Centre for Agricultural Economics and Policy Research (NCAP) joining hand together with the Indian Society of Agricultural Economics (ISAE) organized a national colloquium to steer a debate on the agrarian crisis in India. A group of policy makers, researchers, scientists and experts participated to discuss the results of the studies of various aspects of Indian agriculture and evaluated the current status to provide much needed policy inputs.

The book puts together the selected works in four important relevant thematic sections viz, (1) Indian agriculture: pathways of complacency in quantum jump to deteriorating food security, (2) Intensity and causes of agrarian distress, (3) Opportunities for alleviating agrarian distress, and (4) Strategies to harness opportunities. The final section gives a brief summary of the contributions and collates opinions of various expert groups in a single space, draws meaningful inferences and provides a careful synthesis along with an overview of the state of Indian agriculture as well as the problems looming large. The policy responses to contain the crisis however, require further impetus in the present context. Thus, it is expected that the book will draw concerned policy attention on an urgent manner.

The book is the outcome of the joint National Seminar on Agrarian Distress in India, where a galaxy of experts, senior researchers and policy makers participated and the deliberations are fine-tuned subsequently. We acknowledge and express our gratitude to the society (ISAE) for encouraging, supporting and promoting this activity on the subject of topical importance. We are specifically grateful to Dr. S.S. Johl, then President of the society and other functionaries of the society at whose instance and insistence the activity was performed. We received incredible support and encouragement from Dr. Mangala Rai, former Director General, Indian Council of Agricultural Research to pursue the activity. We are grateful to him. Experts comments and suggestions of the specialists and experts including Dr. S.S. Acharya, Dr. C. Ramasay, Dr. P.G. Chengappa, Dr. P.K. Joshi, Dr. Ramesh Chand, Dr. Prem Vasistha, Dr. Radha Binod Barman, Dr. T.P. Trivedi, helped the authors to revise their papers. We are grateful to them. The book would not have taken the present status without the contributions and cooperation of the authors of the book, who have readily contributed for it. Their insights into the nature and significance have made the book a rich source of information and the magnitude of the problems of agrarian distress in India. We appreciate their effort. We sincerely acknowledge Dr. B.S. Agarwal, for going through and making necessary changes in the initial draft of the papers

Finally, we thank Director, NCAP and scientists other members for providing the desired platform for launching the debate on agrarian distress in India.

B.C. Barah

Contents

Section III
Opportunities for Alleviating Agrarian Distress

Section IV
Strategies to Harness Opportunities

Section V
Policy Imperatives

List of Contributors

B.C. Barah, Principal Scientist (Economics), National Centre for Agricultural Economics and Policy Research (NCAP), DPS Marg, New Delhi-110 012.
barah48@yahoo.com, www.ncap.res.in

Brajesh Jha, Senior Economist, Institute of Economic Growth, Delhi University, New Delhi-110 007. brajesh@iegindia.org

C. Ramasamy Former, Vice Chancellor, Tamilnadu Agricultural University, Coimbatore, Tamil Nadu. cramasamy@gmail.com

D. Suresh Kumar, Associate Professor, CARDS, Tamilnadu Agricultural University, Coimbatore, Tamil Nadu.
rithusuresh@yahoo.com

E. Revathi, CESS, Centre for Economic and Social Sciences (CESS), Nizamia Observatory, Begumpet, Hyderabad.
revathi@cess.ac.in

H.R. Sharma, Head, Department of Agricultural Economics, Himachal Pradesh Krishi Vishwavidyalay (HPKV), Palampur, HP. hannsraj@hillagri.ernet.in

P.G. Chengappa, Vice Chancellor, University of Agricultural Sciences, Hebal, Bangaluru, Karnataka. chengappapg@gmail.com

P. Parthasarthy Rao, Principal Economist, ICRISAT, Patancheru Hyderabad-502 324, AP. parthasarthy@icrisat.org

P. Prudhvikar Reddy, Professor, Centre for Economic and Social Sciences (CESS), Nizamia Observatory, Begumpet, Hyderabad, prudhvikar@cess.ac.in

Pratap S. Birthal, Principal Economist, ICRISAT, Patancheru, Hyderabad-502 324, AP. psbirthal@icrisat.org

R.S. Deshpande, Director, Institute of Social & Economic Change (ISEC), Nagarbhavi, Bangalore, Karnataka.
deshpande@isec.ac.in

Range Gowde, Institute of Social & Economic Change (ISEC), Nagarbhavi, Bangalore, Karnataka. deshpande@isec.ac.in

Rashi Mittal, National Centre for Agricultural Economics and Policy Research (NCAP), DPS Marg, New Delhi-110 012. rashi@ncap.res.in

S. Galab, Professor, Centre for Economic and Social Sciences (CESS), Nizamia Observatory, Begumpet, Hyderabad. sgalab@cess.ac.in

Sant Kumar, Senior Scientist, National Centre for Agricultural Economics and Policy Research (NCAP), DPS Marg, New Delhi-110 012. santkumar@ncap.res.in

Smita Sirohi Professor, Division of Dairy Economics, National Diary Research Institute (NDRI), Karnal, Haryana. smita@ncap.res.in

Suresh Pal Head, Division of Agricultural Economics, IARI, New Delhi-110 01. suresh_eco@iari.res.in

Tushar Pandey, Executive Vice President and Country Head, Strategic Initiatives, YES Bank Limited, 48 Nyaya Marg, Chanakyapuri, New Delhi-110 021. tushar.pandey@yesbank.in

V. Puhazhendi, General Manager, DEAR, NABARD, Bandra Kurla Complex, Bandra (East), Mumbai. (on leave)

Vijayalakshmi Dega, University of Agricultural Sciences, Hebal, Bangaluru, Karnataka.

V. Ratna Reddy, Director, Livelihood & Resource Management Institute, 12-2-417/18, Sarada Nagar, Mehdipatnam, Hyderabad-500 067. vratnareddy@cess.ac.in

Venetia Vickers, Manager-Corporate Education, /Meta-Culture No. 12, Second Floor, 1st Main, Lazar Layout, Frazer Town, Bangalore-560 005, Karnataka.

1
Introduction

B.C. Barah

Agriculture, the most important sector in Indian economy, is presently at a cross road. Although, the share of agriculture in the national GDP has decreased to 18 per cent in 2007, yet the sector continues to provide livelihood support to more than two-thirds of the total population in India. The slow down in performance of the sector during the past couple of decades is a cause of concern in the overall agenda for food and income security. The aggregate productivity has been low, growth of production decelerating and more importantly, the sector is exposed to risk and uncertainty. More than 80 per cent of the farmers, mostly with marginal and small holdings, depend on agriculture as the primary activity and are highly vulnerable to multiple risks. The downtrend in growth of total factor productivity (a measure of contribution of technology) has increased the vulnerability and induced un-sustainability in farm income. At the same time, the escalating cost of production and un-remunerative prices have made farming increasingly unviable. Although, the sector has to gear up to achieve the targeted 4 per cent growth in the 11th Five-Year Plan, there seems to be a silent but widespread unrest and distress at the farm level. The gap between growing aspiration and shrinking achievement in farming has been increasing. While the expectations have grown leaps and bound, the achievements of farm enterprises lagged behind the required pace due to a variety of factors, which have aggravated the state of imbalance. Moreover, the inability to break the yield barrier has accentuated frustration and farmers' distress.

Several factors have contributed to the imbalance, of which the low agricultural productivity and its stagnation are notable. Among the supply side factors, declining factor productivity, degradation of soil and water resources due to intensive agriculture, inequitable access to input and imperfection in output market are prominent. The combined effect of these factors has resulted in escalation of the cost of production, low and uncertain production and falling farm income. The failure of the farming sector to absorb the rapidly growing pool of manpower in rural areas has also caused the decline in household income to add fuel to fire in the distress scenario.

This chapter synthesizes the issues of agrarian crisis in India with a purpose to derive ameliorating measures and policy imperatives to deal with the emergent situation. In particular, the chapter aims to:

- Assess the nature and magnitude of agrarian distress in India,
- Diagnose cause and effect relationship in production stagnation as well as deteriorating food security, and
- Derive appropriate policy interventions and strategies to tackle the issues of agrarian distress.

Agrarian Distress in India: Evidences

The distress among the rural community, allegedly manifested in farmers' suicide, is commonly attributed to debt trap, crop failure and/or yield loss. The agrarian crisis and distress of the rural population which is the outcome of accumulated eventualities and adversity, has emerged as a nation wide phenomenon and a leading policy issue. Decelerating agricultural growth, stagnating yields, erosion of farm income, shrinking access and entitlement to food are the manifestations of agrarian crisis. The increased incidences of farmers' suicides, that have hogged the media limelight, are undoubtedly the most unfortunate reflection of agrarian distress which cannot be brushed aside as an event merely associated with drought or other natural disasters.

Declining Agricultural Output and Decelerating Growth: The macro economic scenario and growth trends show loss of dynamism in agriculture and allied sectors during the past 10-15 years. Growth of agricultural GDP decelerated from over 3.5 per cent per year during 1981-82 to 1996-97 to only around 2 per cent during 1997-98 to 2004-05. The deceleration is observed in almost all major sub-sectors, including horticulture, livestock, and fisheries where, growth was expected to be high. Under the circumstances, achieving targeted growth rate of 4 per cent will be a herculian task as it has been well below the target in the 9th and 10th five year plan period.

The growth rates of production of the important food crops, which are critical to common people, have drastically slowed down over the years, particularly after 2000 (Table 1.1). The growth rate of rice production tumbled down from 3.62 per cent in the 1980s to less than one per cent in the first half of the 2000s. A similar picture also emerged for wheat. The same of coarse cereals like, bajra and millets, which are major food crops of the poor, are still more disappointing.

Table 1.1: Annual Compound Growth Rates of Production of Important Crops in India

(in per cent)

Crop	*1980s*	*1990s*	*2000s*
Rice	3.62	2.02	0.71
Wheat	3.57	3.57	-0.30
Ragi	-0.10	-0.80	-2.68
Small Millets	-3.23	-5.88	-2.63
Barley	-3.48	-0.64	-2.27
Potato	5.17	5.44	-0.50

Source: Based on data from *Agricultural Statistics at a Glance* (various issues)

Another crucial indicator of looming agrarian crisis is the steep decline in the growth of total factor productivity (TFP) during 1990s as compared to the 1980s, indicating the decreasing importance of technological contribution to food production. The TFP of paddy, for instance, has decreased in Punjab and Haryana during the period 1985-2000, remained unchanged in Assam, West Bengal, Karnataka and Uttar Pradesh and increased marginally (1-2 per cent per annum) in Andhra Pradesh, Bihar, Madhya Pradesh and Tamil Nadu (Kumar, 2006). The decadal analysis of the productivity growth of this most important cereal crop has shown that compound growth rate (CGR) of paddy yield after peaking during the 1980s (2.97 per cent), decelerated rapidly since the early 1990s, depicting an inverted bowl shaped growth pattern. The downside growth pattern is a worrisome issue and further continuation of the trend is likely to have deteriorating impact on rural livelihoods (Barah, 2005).

Declining Per Capita Availability of Food: The shrinking availability of food is widely recognized as an important evidence of agrarian crisis and also a cause of social tension that could even culminate in the form of food riots, as have been witnessed in some parts of the globe. The per capita availability of foodgrains has declined from 181kg/annum in 1995 to 154 kg/annum in 2005 (Figure 1.1). The trend is similar for both the important cereal crops of rice and wheat, although the magnitude of decline is more serious in case of rice. The annual per capita availability of rice shows a decline to a level of 64 kg in 2005, which is 16 kg less than 1995 level of 80 kg per capita. Rice is the major staple food besides being the primary activity of the small farmers contributing to household food-security. Due to socio-cultural relevance, biophysical nature of adaptation across diverse ecosystems and geo-politico environment, the cultivation of rice has been a preferred choice of the poor farmers. The declining per capita availability of foodgrains in general, and rice in particular, has profound implications for the household food-security in India.

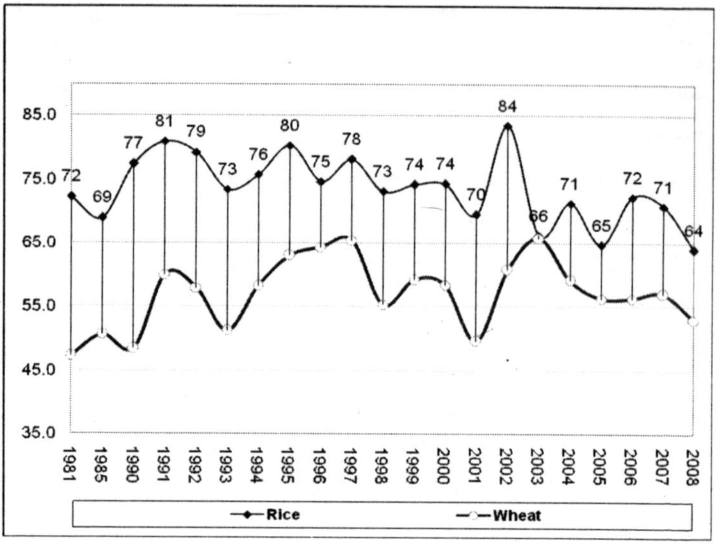

Figure 1.1: Trends in Per Capita Availability of rice and wheat in India (Kg/year)

Source: Based on data from Agricultural Statistics at a Glance

The Graph as in Next Page

Eroding Farm Profitability: In addition to decelerating or stagnating growth trend in production and productivity, the available Figures on returns to farming are the striking evidence of simmering distress in the countryside. Looking at the NSSO and Cost of Cultivation statistics, it is apparent that the returns from agriculture for small and marginal farmers are not adequate to take them above the poverty line, in case agriculture is the only source of income.

In 2002-03, the average return per hectare was Rs. 6,756 in *kharif* and Rs. 9,290 in *rabi* seasons (Sriram, 2008). With nearly 83 per cent of the farming households cultivating an average land size of < 2 ha and 65 per cent cultivating an average land size of < 1.0 ha during *kharif* and *rabi* seasons, respectively, it is obvious that the per capita income from crop farming is very low for the majority of the rural households. Based on the same data, another study has also highlighted that the per capita returns from farming are less than Rs. 8 per day for an average family size of 5.5 members, thus emphasizing that other sources of income are necessary for the farmer household to stay above the poverty line (Mishra, 2007).

Although there are inter-regional variations in farm profitability due to inter crop variations in net returns per hectare, in general, scenario of income generation from crop production is not very encouraging. In five out of six major rice-growing states, in the triennium ending 2003-04, the share of returns to entrepreneurship in value added by the output was negative, which indicates economic unviablility of rice cultivation (Table 1.2). As per more recent estimates (2006), the net returns over cost A2+FL in paddy cultivation ranged from Rs. 1453/ha under rainfed conditions in Bihar to Rs. 15868/ha under irrigated system in Punjab (Table 1.3). The predominantly rice-producing states like Orissa, Bihar, West Bengal, that have low net returns of about Rs. 1500-3000/ha, also have high incidence of poverty. A large majority of farmers owning less than one hectare of land depend on a single crop of rice. The rural poverty line in 2004-05 was about Rs. 358/person/month, which implies that for an average farming family of five members at least Rs. 21,480 are required annually to escape poverty line. Clearly the returns from rice cultivation are not suffice to provide them two square meals daily. It is noteworthy that with the poverty line of Rs. 356.30/person/month at 1973-74 base year in rural areas, a farmer is required to earn at least Rs. 23 from other sources of income to cross the poverty line. Given the escalating input prices, unremunerative output price and remote scope for non-farm opportunity, it depicts the stainability of an average Indian farmer.

Even in the agriculturally well endowed state of Punjab, where crop productivity is highest in the country, the average farm size is 3.79 ha (2.5-times larger than the national average), and the profitability of rice and wheat is positive, yet the economic viability of agriculture is at stake. Overall benefit-cost analysis has revealed that the annual net income from rice and wheat is less than the starting salary of a Grade IV employee in the public offices (Acharya 2007).

Table 1.2: Factor Share in Value–Added Output: Triennium Ending 2003-04

States	*Share in All-India (%)*	*Factor share in value-added to output (%)*			
		Land	*Labour*	*Capital*	*Residual (Profit/Loss)*
		Paddy			
Andhra Pradesh	12.7	30.14	33.25	49.71	-13.11
Tamil Nadu	5.7	22.02	32.87	53.26	-8.15
Orissa	7.5	27.72	44.05	54.51	-26.27
West Bengal	14.5	25.01	50.44	55.46	-30.91
Assam	3.9	23.73	45.35	39.82	-8.90
Punjab	11.1	34.25	12.79	34.47	18.49
		Wheat			
Punjab	20.9	33.08	7.60	31.78	27.54
Uttar Pradesh	34.7	23.58	15.34	33.99	27.09
MP	8.6	25.04	15.58	33.23	26.15
		Cotton			
Andhra Pradesh	11.4	29.06	28.70	49.31	-7.07
Tamil Nadu	1.1	24.77	52.46	40.87	-18.10
Maharashtra	17.1	16.74	33.82	81.57	-32.12
Gujarat	36.6	16.84	40.68	63.44	-20.96
		Sugarcane			
Andhra Pradesh	6.3	18.67	33.89	46.14	1.30
Tamil Nadu	12.5	12.46	33.15	37.91	16.47
Maharashtra	13.8	27.59	32.54	57.81	-17.94
Uttar Pradesh	44.6	39.00	21.31	25.04	14.66
		Groundnut			
Andhra Pradesh	17.1	31.56	38.44	75.32	-45.32
Tamil Nadu	13.8	28.11	51.49	83.2	-62.80
Maharashtra	5.1	12.23	29.96	33.88	23.93
Gujarat	42.4	23.13	21.77	69.53	-14.43

Source: Reddy and Reddy (2007)

In most of the states, the commercial crops like sugarcane, cotton and groundnut have also recorded a decline in the ratio of gross returns over cost C2 in TE 2003-04 compared to TE 1973-74 (Reddy and Reddy, 2007). The case of cotton is particularly notable, where net income over A2 + FL was reported to be negative in Gujarat, Tamil Nadu, Madhya Pradesh and Punjab, while it was very low in other important cotton–producing regions (Ramasamy and Suresh Kumar, 2007).

Table 1.3: Changing Costs of Production of Rice Across Different Ecosystems (A2+FL, Rs/q)

States	*Cost of prod (Rs/q)*			*Yield (t/ha)*			*Net return over A2+ FL (2006)*
	1997-98	*2003-04*	*2006-07*	*1997-98*	*2003-04*	*2006-07*	
			Irrigated system				
Andhra Pradesh	258	350	397	4.38	5.37	5.24	11465
Punjab	210	-	344	5.28	6.51	6.48	15868
Tamilnadu	-	298	509	4.62	4.46	4.46	14340
West Bengal	298	409	465	3.60	3.67	3.30	2621
			Rain fed system				
Bihar	205	367	359	2.14	2.48	2.35	1453
Assam	175	367	440	2.21	2.58	2.90	2892
Karnataka		445	499	2.09	4.58	4.11	9168
Orissa	242	337	409	2.85	3.18	2.35	3225
MadhyaPradesh	304	385	509	2.02	1.85	2.94	2297

High Incidence of Indebtedness: As millions of small farmers have very little disposable production and income, borrowing is unavoidable to carry out the farm activity and to meet the basic household needs. This has become a gradual drag factor pushing the farmers into the deep debt trap. The impact of the phenomenon on the subsistence-oriented system is disastrous. The available literature shows that growing indebtedness has increased the probability of suicides. Radhakrishna Committee on Agricultural Indebtedness (2007) has stated that the credit-related problems of farmers' indebtedness have assumed serious magnitude across the country in the recent years (Table 1.4).

In 2002-03, over four crore farming families were indebted to an average debt of Rs. 12,585 per household, an amount higher than the average annual household farm income. The incidence of indebtedness was very high in Andhra Pradesh, Tamil Nadu, Punjab,

Kerela and Karnataka, where over 60 per cent of the farm households were indebted. In the states of Maharashtra, Gujarat, Rajasthan, Haryana, Madhya Pradesh and West Bengal over half of the farm households were under debt. Except for West Bengal, the average amount of outstanding loan was higher than the all-India average.

Farmers' Suicides: In recent years, the crisis got boosted due to market volatility, increasing incidence of weather-inflicted uncertainties and other systematic flows. The successive droughts specifically aggravated the situation during the decade of 1990s and thereafter. The crisis deepened alarmingly in the distress–stricken major states of Karnataka, Andhra Pradesh and Maharashtra. According to data presented by National Crime Records Bureau (NCRB), the growth rate in suicide mortality rates of farmers during 1996-2005 was 4.8 per cent per year. The male-suicide rate was high in the states of Kerala (195 suicides), Maharashtra (51), Karnataka (41), Andhra Pradesh (33), Tamil Nadu (32) and West Bengal (20), where rural indebtedness was also high. The agrarian distress in these states perhaps amplified to culminate in the spate of suicides by farmers.

Table 1.4: Incidence of Indebtedness in Major States: 2003

States	*Estimated number of indebted farmer households ('00)*	*Indebted farmer households (%)*	*Average loan per household (Rs)*
Andhra Pradesh	49493	82.0	23965
Tamil Nadu	28954	74.5	23963
Punjab	12069	65.4	41576
Kerala	14126	64.4	33907
Karnataka	24897	61.6	18135
Maharashtra	36098	54.8	16973
Haryana	10330	53.1	26007
Rajasthan	27828	52.4	18372
Gujarat	19644	51.9	15526
Madhya Pradesh	32110	50.8	14218
West Bengal	34696	50.1	10931
Orissa	20250	47.8	5871
Uttar Pradesh	69199	40.3	7425
Himachal Pradesh	3030	33.4	9618
Bihar	23383	33.0	4476
Jammu and Kashmir	3003	31.8	1903
Assam	4536	18.1	813
All-India	434242	48.6	12585

Source: National Sample Survey Organization (NSSO), Situation Assessment Survey of Farmers (2003)

Factors Causing Agrarian Distress

Considering the broad shifts in agriculture over the past few decades, there are significant and fundamental changes in the way agricultural practices have emerged. The shortfalls in management of these fundamental changes in some areas have led to agrarian distress. A brief discussion on the important determinants of agrarian crisis in India follows:

Inequitable Penetration of Agricultural Technology: An efficient technology plays a pivotal role in modern agriculture, enhancing productivity and minimizing cost of production. It is well–known that green revolution laid emphasis on high–yielding modern varieties of rice and wheat, which paid high dividend because the dissemination strategy was rightfully accompanied by necessary policy supports and effective land augmenting strategies, input delivery system and market institutions. The synergy of technology and the policy environment worked in perfect harmony, which triggered a comfortable food–security scenario in the country.

Unfortunately, the green revolution concentrated only in the irrigated tracts in the northwest and southern regions. In a stark contrast, the vast and highly diverse rainfed areas accounting for around two-thirds of the total cropped area, were bypassed by the ostracized green revolution.

Ironically, even after more than half a century of agricultural planning, the deprivation and neglect continues in the rainfed areas. The states such as Orissa, Bihar and Karnataka, represent typical rainfed areas, where productivity of rainfed rice is low together with low level of adoption of modern variety and irrigation. On the other hand, in Punjab where 97.5 per cent cropped area is under irrigation, the productivity has more than doubled (Table 1.5).

Table 1.5: Impact of Technology Adoption on Productivity of Rice in Selected States of India

State	*HYVs (%)*	*Irrigation (%)*	*Productivity (t/ha)*
Orissa	64	32	1.79
Bihar	46	59	1.81
Karnataka	75	23	2.68
Punjab	98	97.5	3.69

Source: GoI (2007)

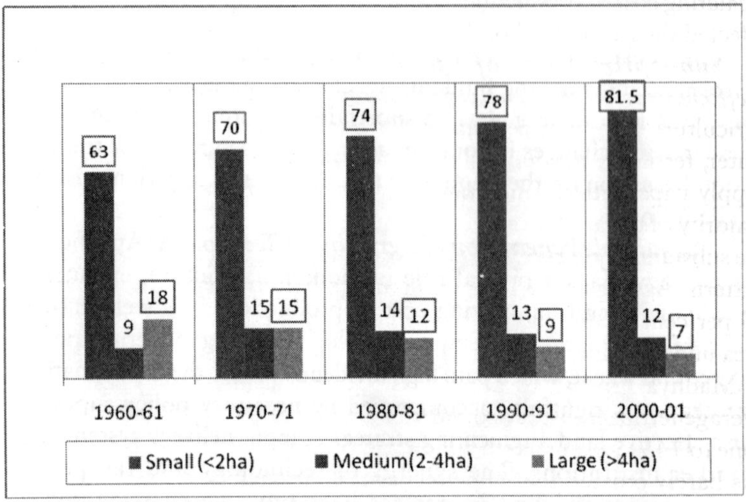

Figure 1.2: Trend in Distribution of Operational Holdings by Farm–Size Group (% of Total)

Although, the rainfed area has huge untapped potential, the region failed to derive benefits of modern technology due to its poor penetration. The unequal treatment in the vast rainfed areas has caused low and uncertain factor productivity in perpetuity, which in turn, resulted in higher and increased incidence of poverty, besides exacerbating regional disparity.

***Declining Farm Size and Small Farmers Orientation of Agriculture*:** The unabated marginalization of landholdings has made farm households more susceptible to volatility in agriculture. The proportion of small farmers has increased from 63 per cent in 1960-61 to 81.5 per cent in 2000-01 (Figure 1.2). During the past four decades, nearly 20 per cent small farmers have been added to the farming sector, while the number of large farmers has decreased to half (Jha, 2001). In the case of area share too, the trend is identical. Research studies have suggested that small farmers are efficient in production, but productivity fluctuations resulting in loss of production and returns to agriculture are the major source of distress among the small farmers, who are unable to manage agricultural risks when they exceed the self-resilient limit. The farmers usually manage the risk, first with their own means, but once the self-adjustment mechanisms are exhausted, crisis becomes inevitable. On the one hand, decline in net return to farming and on the other hand, escalating cost of production at a rate faster than the value of output, affected the rural livelihood and deepened the distress.

Sub–optimal Use of Inputs, Institutional Rigidity and Inefficiencies in Input Delivery System: The input–intensive modern agriculture has created demand for quality seed, timely supply of water, fertilizer, machines, labour and credit far in excess of their supply capabilities. Thus, the input–usages continue to be low in majority of the areas (Ramasamy and Suresh Kumar, 2007) and there are substantial inter–regional and inter–farm variations in input–use pattern. Aggregate average input–usage pattern shows that nearly 60 per cent of the planted area under crops is rainfed. The paddy area under irrigation varies widely, from a low level of 13.3 per cent in Madhya Pradesh to as high as 99.9 per cent in Haryana. The average fertilizer–use is around 96 kg/ha (in range of 35kg/ha in east zone to 143 kg/ha in north zone), and overall seed replacement ratio has barely crossed the 20 per cent mark, though the rate is almost 100 per cent in case of hybrid rice, Bt cotton, etc. It is worth noting that although the modern inputs are indispensable for increasing crop production, their indiscriminate or imbalanced use can adversely affect proper realization of the technological potential.

The institutional rigidity and inefficiencies in delivery system in seeds, credit, agricultural insurance and extension services are the major limiting factors for the securer livelihoods of small farmers. It is ironical that although a huge stock of technology including modern HYVs is available, only a fraction of it has reached the farmers. For instance, of the 720 modern rice varieties developed during the past 4 decades, only a few, such as 'Swarna', developed a quarter century ago was adopted widely covering more than 25 per cent area across the country. Bottlenecks in the efficient distribution of inputs lead to unethical practices of spurious agro–chemicals and seeds, inadvertently affecting production due to which farmers have to incur loses.

Ineffective Pricing Policy: The remunerative product prices and timely access to information are crucial to achieve enhanced income for the resource–poor farmers. However, under the current scenario, pricing policy is ineffective and opportunities for income supplementation are sparse at the household level. The minimum support price (MSP) as a measure to hedge against production and/or income fluctuation, has not reach the poor adequately, as it is implemented in selected crops and in selected regions. The commodities, especially the perishable ones, like tomato, potato, fruits and vegetable, dairy products, fish, etc., which are not supported by minimum guaranteed price are subjected to price crash at the harvest time and hike in the off-season. Since the small farmers practising agricultural diversification are on the edge, therefore, any fluctuation in production and prices adversely affects their farm income. The effect gets further accentuated due to distorted delivery mechanism, lack of policy support and market.

Growing Agricultural Risks: The scenario analysis of risk in agriculture reveals that risk in production is an important source of food insecurity, particularly at the household level and more importantly, the dimension of risk is changing over time and space (Barah, 2007). The decomposition model for rice production showed that in the decade of 1980s, the subsistence producers faced higher yield risk, particularly in the rainfed areas (66 per cent), while irrigated regions experienced lesser yield risk (53 per cent). The irrigated areas were more prone to price risk (69 per cent) that accentuated further to 71 per cent in 1990-03. The technological and policy interventions reduced yield risk in the irrigated areas, but in their absence, risk aggravated in the rainfed areas. In terms of spatial spread of risk, it has been observed that yield risk has not only increased in magnitude over time but has also spread to more districts in the rainfed areas. Over the years, the yield risk decreased in 126 districts but increased in 247 districts, while price risk decreased in 94 districts and increased in 279 districts. It is indeed a cause of concern that despite availability of HYVs, including the resilient varieties to various stresses, and other improved technologies, the production performance has remained below the satisfactory level and of the same time risk has increased over time both in unfavourable environments and significantly well-endowed areas (Barah, 2007). Growing risk directly impinges on the livelihood of the small farmers, whose primary production activity revolves around fulfilling the household consumption need.

Rising Cost of Cultivation: Several factors such as inequitable access to technologies, quality, quantity and timeliness of inputs, fragmentation of ownership holdings, etc. have directly or indirectly led to increase in the cost of cultivation without compensating increase in returns from farm activity. Additionally, Indian agriculture has moved to a stage where almost all the inputs/resources come from the markets, thus increasing the cash component in the cost of cultivation. This is the primary cause of erosion of net income flow to the households. The escalating cost of cultivation induces the farmers to borrow indiscriminately thus, augmenting indebtedness. The crux of the matter is that, farmers resort to heavy borrowing to meet the rising cost of cultivation or to invest on irrigation well and pump sets (Galab and Revathi, 2007) to increase production and income. But, on account of shrinking farm income, they are left with very little surplus and are unable to repay the debt and thus, get trapped into the indebtedness net. The cost-push indebtedness has emerged as the most critical factor of agrarian distress.

In nutshell, the Johl Committee Report has succinctly summarized the gamut of causative factors of agrarian distress (Box 1.1).

Policy Strategy and Redressal Measures

The cycle of intermittent agricultural growth has weakened the livelihood of the vulnerable sections, making them increasingly sensitive to the slightest economic shock. Appropriate interventions are required to boost growth in production and reduce the impact of income volatility. Provision of adequate infrastructure, particularly increased investment on irrigation, and market reforms are the drivers for increasing land productivity and growth.

- The reform agenda in technology development as well as institutional arrangement should aim at improving input–use efficiency to enhance income and ensure livelihood of the farming community. It is required that technology is tuned to location specificity in various ecosystems and is accessible to small farmers. Given the dominance of small farmers in agricultural production system and their rapidly increasing numbers, developing technologies for small farmers should be the top order. Enhancing adoption and adequate access to appropriate technology is indispensable to rescue small farmers from the farming distress confronting them. For the technology to be effective and socially beneficial, a well designed seed system and institutional framework for dissemination, guaranteed remunerative prices and market infrastructure needs to be developed in tandem with the varietal development programmes.
- The existing institutions, such as contract farming, Self-help Groups, Joint Liability Groups, Farmers Club, Farmers' Field Schools, Field Management Committees, etc. should be strengthened to help preparing of the small farmers to face newer challenges and adopt innovations. Well functioning

institutions provide solution to effective implementation of the welfare schemes including credit delivery, facilitating the provision of Kisan Credit Cards (KCC) among the farmers. Programmes on capacity building, training on newer technology and developmental schemes for the farming community are essential to boost adoption.

Box 1.1: Matrix of Issues in Agrarian Distress

Issues	*Demand*	*Supply*
Output/ Price/ Income	• Yield risk because of weather, water and power unavailability, pests, and spurious quality of inputs among others. • Cultivation is not profitable. • Income not sufficient. It is difficult to meet higher education need of wards, medical requirements of family members and other social obligations.	• Increased volatility due to global prices. • Price distortion through subsidies by developed countries. • Low tariff in India. • Minimum support price not always functional. • Futures market—a virtual platform with price volatility being the basis through which hedger/ speculator can operate.
Input	• Supplier-induced-demand is on the rise. This is credit intensive and an important reason for putting the farmer in a quagmire of indebtedness. • There is deskilling. With new technology come new methods of cultivation. Social capital of cultivation know-ledge is rendered redundant. A case is the introduction of *Bacillus thuringiensis* (Bt) cotton. • Greater investments in assets like bore wells in Andhra Pradesh not only increases cost, but has also led to a tragedy when the investments failed.	• No link between publicly funded research and its extension. This is particularly missing for crops/ cultivation in rainfed/dry land areas. • Technological change is sub-stantial and there is an increa-sing reliance on the unregu-lated private suppliers. • Inadequate public investment in agriculture (spread of irrigation in and regions has been a casualty).
Credit	• Formal sources not timely. • Repayment difficult during crop loss and price shocks. • Instead of getting them out of credti, the system draws them into it.	• Formal sources: Decline in the number of branches, decline in agricultural credit/direct finance to agriculture as a percentage of net bank credit, and there is a shift to value addition activities.

Contd.

Issues	*Demand*	*Supply*
	• Difficulties in meeting consumption requirements and other social obligations. • An increase in market induced consumerism.	• Increasing dependence on informal sources-relatively more among smaller farmers.
Other issues	• Political dominance of money lender and/or input dealer and output buyer. • Higher family size: more daughers-greater dowry burden. • Lack of social support.	• Interlinked credit, input and output markets. • Non-farm income opprtunities limited. • Public health response to occupational health hazards of farming is wanting. • Easy availability of pesticides and other hazardous substances.

Source: Reserve Bank of India (2006)

- As risk is an unavoidable feature of Indian farm economy, rather than resorting to isolated fire-fighting measures, tackling the problem of risk-induced agrarian distress requires more concerted efforts. The farmers in India are progressively exposed to multiple sources of risks and hence, there is need for protective policy instruments to hedge against farming risk and ensure their livelihood security. Nonetheless, the responses of the government remain far from effective. Newer products such as *Varsha Bima, Sookha Suraksha Kavach, Flood Insurance, Livestock Insurances, Fruit and Vegetable Insurance, Frost Insurance,* etc. are being developed, but these need to be tested for socio-economic feasibility and adaptability under diverse regional specificities before launching the products on a wider scale (AIC, 2005). Moreover, in order to ensure sustainability, proper blending of public-private partnership is essential in managing and developing business model for insurance products in the long-run. Ensuring livelihood of a large mass of small and marginal farmers, landless labourers and other poor is a public responsibility since the private players will be unwilling to enter into the business of helping the economically marooned farmers. Although the business model for insurance as commercial venture in coffee, tea, horticulture, livestock sectors, etc. has been developed, which has attracted the corporate agencies, successful insurance products should be replicated for other crops. The pricing of insurance product (premium) has to be designed in a pro-farmers domain for achieving societal benefits.

- In addressing the problems of rainfed areas, agricultural diversification is considered important means of income enhancement for improving rural livelihood. More research and development in the hitherto neglected crops such as jowar, bajra, maize and millets, etc. is required for enhancing their production, while keeping the cost of production low in the resource-poor rainfed areas. Also, promoting animal-based income generation activities and developing other non-farm enterprises (Jha, Chapter 9) assume great importance for combating agrarian distress.

Summing Up

The economic condition of the farmers is generally under stress due to drastic decline in farm income, as the cost of production is escalating at a rate faster than the value of output. In addition, the cumulative effect of widely fluctuating productivity and the risks drive the agrarian distress even deeper. As the average net returns from crop cultivation are not adequate to keep the farm households above poverty line, R&D and policy interventions are required to increase output, reduce input cost, improve market facility and enhance non-farm opportunities in the rural areas.

Indian agriculture is basically a small farmer oriented system. Over the years, not only the numbers of small farmers are increasing rapidly, but the size of landholding is also shrinking to a low level. The modern technology, therefore, should be easily accessible to small farmers to derive benefits from it. Technology management and dissemination also require collective action of the farmers group, particularly among the resource-poor farmers in the rainfed areas. The groups, rather than individual, should be entrusted with the responsibility of agricultural input management and product marketing in the rural areas.

Past experience has shown that due to lack of synergy between technology and supportive policy, the technology solution had met with a partial success. That is why although the farmers adopt modern variety, the crop productivity has stagnated due to non-targeted variety. In view of rapidly changing knowledge-intensive agricultural technology, the need for proactive agricultural

R&D system as well as continuous capacity building through innovative developmental programme on enhancing human capital has become imperative. The application of ICT to provide generic information to empower farmers with scientific know-how, market intelligence and prices is critical.

The farming community has become the victim of cost-push indebtedness. The cost of production is increasing steeply due to intensive use of external inputs, supply side bottleneck and inefficient pricing policy. Further, over-exploitation and degradation of natural resources has led to unsustainable agricultural production which in turn, has impinged on farm profitability. The elements of the multi-pronged policy strategy to tackle the problems of farm profitability and rural indebtedness include: institutional reforms for efficient input delivery and output sale in rural areas and oobjective assessment of farmers risk for developing alternative farmer-friendly insurance products to hedge against risk. There is ardent requirement of a carefully designed credit delivery system for improving credit inclusion and providing adequate counselling to the borrowers by introduction of credit management advisory services. The social isolation of the indebted farmers and burden of insult due to indebtedness are depressing and need to be addressed with collective efforts of Social Advisory Counselling Services in the rural areas. The rendering of advisory services would address the problems of the distress-ridden farmers and improve social support system. Its delivery mechanism could be under the aegis of ongoing campaign for rural communication through ICT, Kisan call centres, community managed radio station for a particular area and other media, which are expected to pay large dividend in alleviating farmers' distress.

References

Acharya, S.S. (2007), *India's Current Agrarian Distress: Intensity, and Causes*, Chairperson's observations in the Technical Session II in the *National Seminar on Agrarian Distress: Causes and Intensity*, National Centre for Agricultural Economics and Policy Research (NCAP), New Delhi.

AIC (Agriculture Insurance Co) (2005), *Varsha Bima*, New Delhi.

Barah, B.C. (2005), *Dynamics of Rice Production System in India*, Occasional Paper No. 47, NABARD, Mumbai.

Barah, B.C. (2007), Changing pattern of agricultural risk: An analysis of rice risk, *National Seminar on Agrarian Distress: Causes and Remedies*,

National Centre for Agricultural Economics and Policy Research (NCAP), New Delhi.

Galab, S. and E. Revathi (2007), Agrarian distress in Andhra Pradesh: Causes and remedies, *National Seminar on Agrarian Distress: Causes and Remedies*, National Centre for Agricultural Economics and Policy Research (NCAP), New Delhi.

GoI (Government of India) (2007), *Agricultural Statistics at a Glance*, Ministry of Agriculture, New Delhi.

Jha, B. (2007), Enhancong income through non-farm employment, *National Seminar on Agrarian Distress: Causes and Remedies*, National Centre for Agricultural Economics and Policy Research (NCAP), New Delhi.

Jha, D. (2001), Agricultural research and small farms, Presidential Address, *Indian Journal of Agricultural Economics*, **56** (1).

Kumar, P. (2006), Total factor productivity in India, *Agricultural Economics Research Review*, **21** (Conference Issue).

Mishra, Srijit (2007), *Risks, Farmers' Suicides and Agrarian Crisis in India: Is There a Way Out?* Working Paper WP-2007-014, Indira Gandhi Institute of Development Research, Mumbai

Ramasamy, C. and D. Suresh, Kumar (2007), Regional difference in farm profitability in Indian agriculture, *National Seminar on Agrarian Distress: Causes and Remedies*, National Centre for Agricultural Economics and Policy Research, New Delhi.

RBI (Reserve Bank of India) (2006), *Report of the Working Group to Suggest Measures to Assist Distressed Farmers*, Mumbai (Chairperson: S. S. Johl).

Reddy, V. Ratna and Prudhvikar Reddy (2007) Agrarian crisis and labour market dilemma in India, *National Seminar on Agrarian Distress: Causes and Remedies*, National Centre for Agricultural Economics and Policy Research (NCAP), New Delhi.

Sriram, M.S. (2008), Agrarian distress and rural credit: Peeling the onion. In: *State of India's Livelihoods*, Eds: Sankar Datta and Vipin sharma, Access Development Services, New Delhi, Chap. 8.

Section I

PROBLEMS IN INDIAN AGRICULTURE: CAUSES OF AGRARIAN DISTRESS

Agriculture has been a way of economic life and main source of livelihood for the vast majority of households in rural India. Despite a decline in the share of agricultural sector in GDP to less than one-fifth at present from over one-half at the. time of Independence, it remains the predominant employer for more than 50 percent of the rural workforce in country. During the past five and the half decades of planned economic development in India, the sector has come a long way from foodgrain production level of only 51 million tonnes in 1950-51 to 233 million tonnes in 2008-09. Production of cotton, oilseeds and sugarcane has also increased more than four-fold over the period. The increase in production has been brought about largely by productivity enhancement of about 255 per cent in foodgrains, 182 per cent in sugarcane, 142 per cent in oilseeds and 57 per cent in cotton—thanks to a favorable interplay of technology and infrastructural development.

The progress of Indian agriculture has, however, not been consistent over time and can be classified into four distinct phases. The first phase stretching from independence to mid-1960s, emphasized on consolidation and organization of agriculture sector. Development was spearheaded through industrial front and it was expected to have a spread effect on agriculture. The increase in agriculture production at the annual rate of about 3 per cent was dominated .by growth in non-foodgrains. A slower increase in foodgrain production came about due to shift in cropping pattern in favour of superior cereals (wheat and rice), particularly in the better endowed regions. The lack of emphasis on technological change during this phase culminated in extreme food scarcity in mid-1960s.

The advent of new technology changed the situation dramatically in the second phase, spanning mid-1960s to the decade of 1970s. The growth rate of foodgrains was impressive and it came about due to improvement in yield. From the situation of acute shortages at the beginning of the phase, the country surged ahead in achieving self-sufficiency in foodgrain production. Besides the new technology, the strengthening of the institutional backup also contributed to productivity growth, and the transformation in agrarian structure was an important component of agriculture development in this phase.

The growth rate of foodgrain production was further enhanced during the decade of 1980s and in addition, the non-foodgrain production also picked up. It was during this third phase that horticultural crops, floriculture, sericulture and other crop enterprises received significant attention of the policymakers and farmers. The growth rate of agriculture production touched an all-time high of 3.7 per cent. In the phase of near stagnancy in area expansion, the productivity growth about 3 per cent brought about high agricultural growth.

The fourth phase, from the beginning of 1990s, has been marked by considerable slacking of agricultural growth. The seed-water-fertilizer technology that had been the harbinger of growth in agriculture also had its negative externalities to deal with, Additionally, the issues of *its* impact on regional *and* inter-personal inequalities turned important as the disparities in growth performance across crops, regions and farm-sizes became profound. The technology was termed as scale neutral but it was not access neutral. It was argued that all the components of technology were not easily accessible to the marginal and small farmers. The cost of cultivation scaled up on account of escalating expenditure on modern inputs and consequently eroding the farm profitability. All these factors increased vulnerability and risk in Agriculture.

This slow down has been a cause of grave concern as the overall growth of the economy is sensitive to the fluctuations in agriculture sector. In order *to* achieve the targeted *4* per cent growth in agriculture, it is imperative to diagnose the problems afflicting the Indian agriculture sector, ft is with this purpose that the various Chapters in the section I focus on various aspects of the Indian agriculture, like inter-regional and inter-crop variation in profitability of agriculture, trend in agriculture labour markets, access to land and input resources, and changing pattern of agricultural risk etc. In understanding the genesis of the slackening performance of agriculture, the main issue that has been examined, is whether the problems are endogenous or exogenous to the agriculture sector.

2

Regional Differences in Farm Profitability in Indian Agriculture

C. Ramasamy and D. Suresh Kumar

Introduction

The agricultural growth in India has shown considerable inter-regional variations, though in general, there has been a decline in growth of the agriculture sector since 1990 as compared to 1980s. The existence of wide variations in yield, production, and input-use in agriculture and agricultural growth across regions has been confirmed by several studies (Sawant, 1997; Singh *et al.*, 1997; Sardana *et al.*, 1997; Rao and Gopalappa, 2004; Sidhu and Bhullar, 2005; Mathur *et al.*, 2006). Across states, agricultural growth has varied from 28 per cent to–19 per cent during the triennium ending (TE) 1996 to TE 2003. The growing imbalances in agricultural growth and development have led to wide disparity in the status of farmers across regions. A significant proportion of farm households have been trapped in indebtedness. Recent estimates have shown that nearly 49 per cent of farm households are indebted and the average amount per indebted household is about Rs. 12,500/-.

Though a lot of work has been carried out on the nature, pattern and determinants of inter-regional variations in yield, production and agricultural growth, the issues of regional differences in farm profitability and consequent distress of farmers have been addressed to a limited extent. In the light of this observation, the present Chapter has analysed the regional differences in farm profitability in Indian agriculture.

Variations in Resources, Technology and Institutions

Resource Endowments

Land and availability of water for irrigation are the two critical resources for agriculture. While access to land is pre-requisite for crop cultivation, access to irrigation has a significant bearing on the productivity and revenue from crop production.

Landholdings: The agrarian structure in India has witnessed intensification of marginal and small holdings. The marginal and small farmers accounted for 78.2 per cent holdings and 32.4 per cent operated area in 1990-91. In a decade, by 2000-01, the share of this group went up to 81.9 per cent and 39 per cent, in terms of number of holdings and area operated, respectively. This vulnerable picture, evident from the large number of marginal and small farmers operating less than two-fifths of land area implies that the resource-poor farmers still continue to dominate the Indian agriculture (Table 2.1).

Table 2.1: Area Operated by Different Categories of Holdings: All India: 1990–91 To 2000-01

No. of holdings: ('000 No.)
Area operated: ('000 ha)

Category of holdings	*No. of operational holdings*			*Area operated*		
	1990-91	*1995-96*	*2000-01*	*1990-91*	*1995-96*	*2000-01*
Marginal	63389	71179	76122	24894	28121	30088
(less than 1 ha)	(59.4)	(61.6)	(63.0)	(15.0)	(17.2)	(18.82)
Small	20092	21643	22814	28827	30722	32260
(1 to 2 ha)	(18.8)	(18.70)	(18.9)	(17.4)	(18.80)	(20.18)
Semi-medium	13923	14261	14087	38375	38953	38305
(2 to 4 ha)	(13.1)	(12.3)	(11.7)	(23.2)	(23.8)	(23.96)
Medium	7580	7092	6568	44752	41398	38125
(4 to 10 ha)	(7.1)	(6.10)	(5.4)	(27.1)	(25.30)	(23.84)
Large	1654	1404	1230	28659	24163	21124
(>10 ha)	(1.6)	(1.20)	(1.02)	(17.3)	(14.80)	(13.21)
All holdings	106637	115580	120822	165507	163357	159903
	(100.0)	(100.0)	(100.0)	(100.0)	(100.0)	(100.0)

Note: Figures within parentheses indicate the percentages of respective column to total.
2000-01 figures are excluding Jharkhand

Source: Agricultural Statistics at a Glance, 2006, Directorate of Economics and Statistics, Ministry of Agriculture, Government of India, New Delhi.

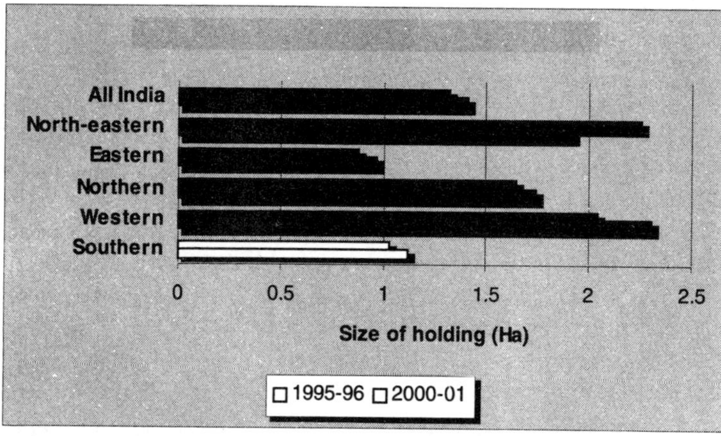

Figure 2.1: Average Size of Holding Across Regions

The area operated by different categories of farmers and the average size of operational holding significantly varied across regions, ranging from 0.88 ha in the eastern region (Bihar, Jharkand, Orissa and West Bengal) to 2.25 ha in north-east region (Assam, Arunachal Pradesh, Mizoram, Nagaland, Meghalaya, Manipur, Sikkim and Tripura) with all-India average of 1.32 ha in 2000-01 (Figure 2.1 & Table 2.2). As the average size of holdings in the eastern region falls under the marginal category, it will be difficult to achieve economies of size in terms of adoption of technologies and input-use. The farmers in eastern region may be forced to work for wage employment for their livelihood as the average size of holdings is less than one ha of land (i.e. marginal holding). Despite relatively larger size of landholdings in the north-east region (i.e. semi-medium holding), this may not provide assured income and employment for their livelihood. As for in the other regions, the average size of holdings is higher than the national average in the western and northern, regions, while in the southern region it is marginally lower.

Irrigation: The positive impact of irrigation on farm profitability is evident from the estimates of relative returns on irrigated and unirrigated holdings. An irrigated farmer in India, on an average, has been found to obtain about Rs. 14,000/ha of cultivated area (at 1992-93 price), whereas the average rainfed farmer got about Rs. 6,000/ha only (Dhawan, 2000).

The percentage of area irrigated by all crops accounted for 40.2 per cent in 2002-03, which varied across states and regions (Table 2.3). Across the states, it varied from 5.5 per cent in Assam to 97.8 per cent in Punjab. Other than Punjab, very few states, Haryana (86.2 per cent),

Table 2.2: Region-wise Area Operated by Different Categories of Holdings in 2000-01

('000 ha)

Region	States	Marginal	Small	Semi-medium	Medium	Large	All holdings	Average size of holding (ha)	
								1995-96	2000-01
Southern	Andhra Pradesh	3104	3565	3795	2855	1080	14400	1.36	1.25
	Karnataka	1492	2742	3429	3317	1327	12307	1.95	1.74
	Kerala	883	300	191	85	112	1569	0.27	0.24
	Tamil Nadu	2159	1712	1551	1094	455	6972	0.91	0.89
	Sub-total	7638	8319	8966	7351	2974	35248	1.12	1.03
Western	Maharashtra	3026	5277	6204	4785	1467	20760	1.87	1.57
	Gujarat	649	1822	2883	3327	1031	9711	2.62	2.35
	Goa	17	8	8	8	12	54	0.84	0.84
	Rajasthan	894	1744	3416	6811	8386	21251	3.96	3.65
	Madhya Pradesh	1398	2828	4122	5448	2576	16372	2.28	2.22
	Chattisgarh	776	1018	1375	1410	643	5223		1.60
	Sub-total	6760	12697	18008	21789	14115	73371	2.31	2.04
Northern	Uttar Pradesh	6648	4366	3906	2580	484	17983	0.86	0.83
	Uttarakhand	243	221	212	132	36	843		0.95
	Punjab	78	242	876	1731	1096	4022	3.79	4.03
	Haryana	317	422	781	1210	821	3550	2.13	2.32
	Jammu and Kashmir	429	250	202	69	12	962	0.76	0.67
	Himachal Pradesh	252	245	243	176	63	979	1.16	1.07
	Sub-total	7967	5746	6220	5898	2512	28339	1.74	1.65
Eastern	Bihar and Jharkand	2907	1296	1544	861	140	6747	0.75	0.58

Contd.

	West Bengal	2759	1607	784	178	219	5547	0.85	0.82
	Orissa	1155	1544	1344	818	220	5081	1.30	1.25
	Sub-total	6821	4447	3672	1857	579	17375	0.97	0.88
North-eastern	Assam	607	718	1001	525	264	3115	1.17	1.20
	Arunachal Pradesh	7	27	97	171	92	394	3.31	3.69
	Mizoram	22	36	29	6	1	93	1.29	1.24
	Nagaland	2	11	72	420	543	1047	4.82	7.28
	Meghalaya	63	83	91	36	5	278	1.33	1.30
	Manipur	40	63	55	14	0	172	1.22	1.15
	Sikkim	12	17	21	20	12	80	1.65	1.57
	Tripura	126	75	43	9	16	270	0.60	0.56
	Sub-total	879	1030	1409	1201	933	5449	1.92	2.25
India		30088	32260	38305	38125	21124	159903	1.41	1.32

Source: *Agricultural Statistics at a Glance* (2006), Directorate of Economics and Statistics, Ministry of Agriculture, Government of India, New Delhi.

Table 2.3: Percentage of Area Covered Under Irrigation Across Regions During 2002-03

Region	States	Percentage of area under irrigation					
		Rice	Sugar-cane	Ground-dnut	Cotton	Red-gram	All crops
Southern	Andhra Pradesh	95.3	92.7	15.2	20.1	1.4	39.2
	Karnataka	70.5	100.0	35.3	14.7	2.0	24.5
	Kerala	58.5	100.0				14.5
	Tamil Nadu	90.7	100.0	31.2	33.9	1.4	50.5
Western	Maharashtra	29.1	100.0	31.0	4.1	1.8	18.1
	Gujarat	54.1	100.0	7.0	41.1	14.0	31.4
	Rajasthan	50.0	96.8	49.0	95.0	3.6	39.9
	Madhya Pradesh	13.3	99.7	8.3	31.2	1.0	25.6
Northern	Uttar Pradesh	66.2	88.8	0.6	92.6	12.3	70.3
	Punjab	99.2	94.6	41.8	99.6	90.5	97.8
	Haryana	99.9	98.7	90.2	98.7	74.8	86.2
	Jammu and	93.1	35.9		56.8		40.3
	Himachal Pradesh	62.5	48.5	7.8		0.8	18.8
Eastern	Bihar	54.6	25.2			0.5	57.5
	West Bengal	42.1	35.7				36.7
	Orissa	29.6	100.0	17.2		0.6	21.8
North-eastern	Assam	8.2					5.5
	Arunachal Pradesh	36.1					16.3
	Mizoram	27.5	7.3			9.3	11.0
	Nagaland	42.8					22.0
	Meghalaya	49.5					26.6
	Manipur	46.1					34.2
	Sikkim	100.0					13.6
	Tripura	18.7		55.0			14.1
India		50.2	91.3	16.5	33.1	4.6	40.2

Source: Agricultural Statistics at a Glance (2006), Directorate of Economics and Statistics, Ministry of Agriculture, Government of India, New Delhi.

**Estimated Triennium average ending from 2000-01 to 2002-03

Uttar Pradesh (70.3 per cent), Bihar (57.5 per cent), Tamil Nadu (50.5 per cent) and Jammi and Kashmir (40.3 per cent), have the irrigated area above the national average.

At the regional level, the northern region has high level of irrigated area and consequently, it is relatively well-developed in terms of agriculture, while the north-eastern region has poor irrigation endowment.

The intra-regional and inter-crop variations have also been found significant. For instance, the percentage of area under irrigation in southern region varies from 14.5 per cent in Kerala to 50.5 per cent in Tamil Nadu. Similar situation has been visualized in almost all the regions. The crop-wise irrigation statistics shows that the percentage of irrigated area is 50.2 per cent under rice, 91.3 per cent in sugarcane, 16.5 per cent in groundnut, 33.1 per cent in cotton and 4.6 per cent in red gram. Like the total irrigated area, the area irrigated under these crops also exhibits inter-regional differences. It is interesting to see that the area irrigated by rice ranged from 8.2 per cent in Assam to 99.99 per cent in Haryana. Though rice is a water-intensive crop which consumes 13000 M^3 of water per hectare, it is also grown as semi-dry or rainfed crop in many states and regions of the country.

Technology: Modern high-yielding variety seeds, fertilizer-use, and plant protection assume importance in accelerating agricultural production. The increasing demand for food and non-food agricultural commodities coupled with various production constraints have necessitated the evolution of new technologies and their adoption across regions. Thus, it is crucial to look at the technology adoption and its contribution for productivity and production enhancement across the regions.

Modern Varieties: Despite almost three-fourths of the gross cropped area in the country being dominated by food crops, mainly cereals and pulses, adoption of modern varieties of these crops is not at the desired level (Ramasamy, 2003). The spread of modern varieties across regions exhibits wide variations both, spatially and temporally (Table 2.4). The total area under modern varieties increased from 30 per cent during 1970s to 71.8 per cent during 1990s. The spread of modern varieties varied from about 71 per cent in the western and eastern regions to 88 per cent in the southern zone. Among the crops, the area under HYVs of crops ranged between 2 and 69 per cent across the states. Estimated Gini coefficient of 0.60 implies that there is a wide variation among the states in adoption of HYV crops (Ramasamy and Selvaraj, 2001).

Table 2.4: Spread of Modern Varieties Across Regions

(per cent)

Regions	*1970s*	*1990s*
South zone	57.25	88.15
North Zone	57.10	80.36
East zone	20.53	71.08
West zone	29.55	71.97
India	30.70	71.76

Source: Ramasamy and Selvaraj (2001)

Fertilizers and Pesticides: The consumption of chemical fertilizers across regions has shown a similar pattern as variations in irrigation. On an average, the fertilizer consumption is 88.09 kg/ha in India (Table 2.5) and this varied across regions from 1.74 kg/ha in Nagaland (north-eastern region) to 168.21 kg/ha in Punjab (northern region). Most of the states in southern and northern regions consume fertilizers at a level above the national average. It has been found that both the irrigated area and fertilizer consumption are high in the southern and northern regions.

Table 2.5: Fertilizer Consumption Across Regions During 2002-03

Region	*States*	*Fertilizer consumption (kg/ha)**	*Pesticide consumption (in tonnes tech.grade**)*
Southern	Andhra Pradesh	144.39	4000
	Karnataka	100.77	2020
	Kerala	62.03	754
	Tamil Nadu	135.49	1668
Western	Maharashtra	75.17	3239
	Gujarat	83.48	2822
	Rajasthan	33.53	3040
	Madhya Pradesh	39.67	871
Northern	Uttar Pradesh	123.39	7023
	Punjab	168.21	7005
	Haryana	153.68	5025
	Jammu and Kashmir	60.99	01
	Himachal Pradesh	40.17	302
Eastern	Bihar	90.93	853
	West Bengal	122.77	3250
	Orissa	38.41	1006
North-eastern	Assam	38.69	245
	Arunachal Pradesh	2.67	30
	Mizoram	16.60	08
	Nagaland	1.74	08
	Meghalaya	15.93	06
	Manipur	114.88	20
	Sikkim	9.54	04
	Tripura	24.84	11
India		88.09	43584

Source: Agricultural Statistics at a Glance (2006), Directorate of Economics and Statistics, Ministry of Agriculture, Government of India, New Delhi.

*Estimated triennium average ending from 2000-01 to 2002-03

** Figures pertain to the year 2000-01

Similarly, the consumption of plant protection chemicals, i.e. pesticide has shown significant variations across states and regions. Use of pesticides varied from 1.00 t/ha of technical grade in Jammu and Kashmir (northern) to 7023 t/ha of technical grade in Uttar Pradesh (norhern). It is interesting to note that the pesticide consumption is significantly high in most of the southern and northern states.

Institutional Support

The institutional factors such as government expenditure on agricultural development, flow of institutional credit, and other supports have direct bearing on the agricultural production in any region.

Government Expenditure: Government investment is one of the crucial factors determining agricultural growth in a region. The ratio of agriculture to government expenditure for major states has been worked out to be 22.3 per cent and it varied significantly across states with 46.11 per cent in Punjab to 10.93 per cent in Madhya Pradesh. It is clear that states endowed with more government expenditure show better growth in the agricultural sector (Mathur *et al.*, 2006).

Agricultural Credit: Agricultural credit has been a significant factor in promoting new technology and ushering green revolution in the country. Both central and state governments have taken several initiatives to increase the flow of agricultural credit by expanding functional and geographical coverage of the rural financial institutions and setting up new institutions to cater to the needs of rural areas and weaker sections. Consequently, the use of institutional credit has improved in the agriculture sector across regions. Sidhu and Gill (2006) have found that there are large variations in the availability of institutional credit across states and regions (Table 2.6). While flow of institutional credit per hectare of gross cropped area is high in most of the southern states (Rs. 9403 in Tamil Nadu, Rs. 7666 in Kerala, and Rs. 4604 in Andhra Pradesh) and Punjab (Rs. 5352) in the northern region, the availability of institutional credit is very low in Assam (Rs. 311) in the north-east, Orissa (Rs. 479) in east and Rajasthan (Rs. 667) in western India. Very high use of credit per hectare in southern region and Punjab is attributable to high level of input-use and private capital formation on the farms supported by the lending institutions. Contrarily, the poor growth and availability of institutional agricultural credit in the eastern and north-eastern regions, may partly be responsible for low input-use and low private capital formation on farms. Broadly, the distribution of credit is skewed towards the agriculturally developed regions.

Table 2.6: Distribution of Flow of Institutional Agricultural Credit Across Regions in 2001-02

Region	*States*	*Agricultural credit*		
		Rs crores	*Percentage*	*Rs/ha of GCA*
Southern	Andhra Pradesh	5587	13.5	4604
	Karnataka	4014	9.7	3432
	Kerala	2276	5.5	7666
	TamilNadu	6166	14.9	9403
Northern	Uttar Pradesh+ Uttranchal	4056	9.8	1529
	Punjab	4304	10.4	5352
	Haryana	1821	4.4	2964
	Himachal Pradesh	248	0.6	2555
	Jammu and Kashmir	83	0.2	764
Western	Gujarat	2980	7.2	2809
	Maharashtra	2938	7.1	1352
	Rajasthan	1490	3.6	667
	Madhya Pradesh+ Chattisgarh	1821	4.4	698
Eastern	Orissa	414	1	479
	West Bengal	1573	3.8	1708
	Bihar+Jharkhand	1076	2.6	1075
North-Eastern	Assam	124	0.3	311
All-India		41385	100	2169

Source: Sidhu and Gill (2006)

Soil Testing Laboratories: Adequate institutional support for dissemination and adoption of technologies is crucial to increase productivity and production of crops. The Soil Testing Laboratories (STLs) are one such institutional arrangement aimed at testing soil samples and recommend nutrient requirements for different crops under different agro-climatic conditions. The details of established soil testing laboratories have shown that there is a wide variation across states and regions. Of the total number of 462 STLs in the country, nearly 60 per cent are located in the southern and northern regions, which are agriculturally developed (Table 2.7). The number of STLs is very low in the eastern and north-eastern regions, implying that these regions need to be equipped with more number of such laboratories.

Table 2.7: Details of Soil Testing Laboratories Across Regions

Regions	*Static*	*Mobile*	*Total*
Southern	77	30	107 (23.16)
Western	74	23	97 (20.99)
Northern	144	31	175 (37.87)
Eastern	48	6	54 (11.68)
North-eastern	17	8	25 (5.41)
All-India	362	100	462 (100.00)

Note: Figures within the brackets show percentages in all-India level

Variations in Production and Productivity

Differences in Level

With wide variations in agro-economic conditions and socio-economic diversity, the issue on regional differences in production and productivity of different crops is of crucial interest for effective regional planning. The regional differences in area, production and productivity of rice, sugarcane, cotton, groundnut and red gram have been presented in Table 2.8.

Rice: It is evident from Table 2.8 that all the five regions of rice-growing areas in the country have distinct behavioural pattern and production conditions, which have implications on the overall production. The southern region with about 15.76 per cent of total rice area in the country had contributed 21.84 per cent of the total production, that is, a ratio of production to area share of 1.4 in TE 2004-06. This implies that the unit area has contributed almost one and half times the total production. The eastern region, on the other hand, with 31.86 per cent area had produced approximately 29 per cent of the total output of rice, the ratio being 0.09. Though, rice is the major crop in eastern India (in many cases it is the only crop in the *kharif season*), its productivity performance has been dismal. The corresponding figure in the western region is 6.09 per cent of the total production in 9.03 per cent area. Thus, it has been inferred that the resource endowment directly influences the production performance. The production condition is generally homogeneous under the irrigated situation (as in northern and southern regions), whereas in the eastern India rice is grown under highly diverse ecosystems dominated by rainfed system and almost in natural conditions. The differences in the production environments have been the primary source of variations in production across regions.

Table 2.8: Proportions of Area, Production and Yield of Major Crops Across Regions in India

Crops	Southern region		Western region		Northern region		Eastern region		North-eastern region		All-India	
	1991-93	2004-06	1991-93	2004-06	1991-93	2004-06	1991-93	2004-06	1991-93	2004 06	1991-93	2004-06
Rice												
Area (%)	18.10	15.76	17.14	9.03	19.39	21.76	35.88	31.86	5.93	5.72	100.00	100.00
Production (%)	26.31	21.84	11.49	6.09	24.87	27.74	29.79	28.70	4.40	4.14	100.00	100.00
Yield (t/ha)	2.42	2.68	1.31	1.49	2.61	2.91	1.41	1.71	1.29	1.49	1.75	2.05
Sugarcane												
Area (%)	18.56	17.56	15.00	16.61	57.76	58.95	3.86	2.58	–	–	100.00	100.00
Production (%)	24.74	23.17	18.97	16.96	49.82	52.83	2.89	1.68	–	–	100.00	100.00
Yield (t/ha)	86.09	85.11	83.86	68.66	56.21	58.97	48.89	41.09	–	–	65.10	63.56
Cotton												
Area (%)	20.97	18.57	62.64	67.69	15.81	12.94	–	–	–	–	100.00	100.00
Production (%)	27.45	17.07	44.36	59.61	27.39	22.46	–	–	–	–	100.00	100.00
Yield (kg/ha)	357.33	275.67	277.33	294.83	481.0	573.83	–	–	–	–	280.67	329.67
Groundnut												
Area (%)	56.94	50.93	28.28	38.18	1.58	1.44	3.44	1.34	–	–	100.00	100.00
Production (%)	60.09	34.52	22.67	50.96	1.53	0.96	5.02	1.35	–	–	100.00	100.00
Yield (t/ha)	1.02	1.03	0.84	1.45	0.91	0.77	1.26	0.84	–	–	0.92	1.19
Redgram												
Area (%)	25.38	31.07	51.14	46.99	14.11	10.73	6.34	4.80	–	–	100.00	100.00
Production (%)	12.98	23.46	49.67	49.46	24.56	15.68	8.86	5.63	–	–	100.00	100.00
Yield (kg/ha)	386.0	541.44	698.0	805.22	1113.67	1023.33	982.67	972.50	–	–	637.67	701.33

In addition to production, the differences in productivity performance have also been found significant across the regions. The southern and northern regions had productivity level above the national average. Though the eastern region is rice dominated agriculture, the region had productivity level below the national average of 2.05 t/ha in TE 2004-06.

Sugarcane: It is extensively grown in the northern region, which accounted for 58.95 per cent of area and 52.83 per cent of total sugarcane production in the country. The other major sugarcane-growing regions are southern and western regions. Southern region produced about 23 per cent of total sugarcane production in 18 per cent of sugarcane area of the country. This has been mainly attributed to higher productivity in this region. The western region has shown an interesting picture with about 16 per cent of area producing the same proportion of country's production. There are vast differences in productivity of sugarcane. The productivity of sugarcane in southern and western regions has been found above the national average of 64 t/ha, whereas the picture has been observed little miserable in the northern and eastern regions, where the productivity is very low in comparison to the national average.

Cotton: The western region, leading producer of cotton, accounted for about 68 per cent of total cotton area with a production of nearly 60 per cent of the total cotton production in the country in TE 2004-06. The southern and northern regions together accounted for about 32 per cent of the cotton area, producing nearly 40 per cent of the total cotton production.

Groundnut: It is an important oilseed crop in the country and is produced in almost all the regions, except the north-eastern region. Southern region with about 51 per cent of groundnut area produced nearly 35 per cent of total groundnut production in the country during TE 2004-06. It was followed by the western region, which produced about 51 per cent of groundnut with 38 per cent of the area. Also, the average productivity of the region (1.45 t/ha) was higher than the national average of 1.19 t/ha during this period.

Red gram: The western and southern regions are the major red gram producing regions of the country. Both the regions accounted for about 78 per cent of the total red gram area and produced 74 per cent of total production in the country. Though the northern region produced only 16 per cent of red gram from 10 per cent of the area, it stood first in terms of productivity with 1023 kg/ha yield.

Differences in Growth

The rate of change in area, production and yield of major crops during the period 1990-91 to 2005-06 has been found very disappointing, although it has shown wide regional disparities, particularly in the case of rice, which is the important staple food in the country (Table 2.9).

India registered the annual CGR of 0.1 per cent, 1.07 per cent and 0.97 per cent in area, production and yield of rice for the period 1990-91 to 2005-06. The state-wise analysis has shown that most of the states registered negative growth in area under rice. The growth rate in area varied from-9.48 per cent in Madhya Pradesh (Western region) to 3.43 per cent in Haryana (Northern region). Similarly, the growth in production varied across regions and states and ranged from-11.67 per cent in Madhya Pradesh to 3.68 per cent in Haryana. It is interesting to note that unlike in area and production, the growth in yield has exhibited somewhat different scenario. The negative growth in yield was observed only in three states, viz. Orissa (Eastern region), Madhya Pradesh (Western region) and Tamil Nadu (Southern region). Thus, the positive growth in yield in most of the states compensated for the declining area under rice to increase the total rice production. The dominating yield effect could be attributed to high level of input use, modern varieties, technology and agro-climatic factors. The Northern region, comprising states like, Uttar Pradesh, Punjab and Haryana, has registered high positive growths in area, production and yield of rice when compared to other regions.

The two major commercial crops of India, sugarcane and cotton, have shown declining trends in yield. However, despite negative growth in yield, the positive area effect had led to the marginal growth in production, albeit differentially across regions. The yield of sugarcane has registered a negative growth in almost all the regions and varied from–0.002 per cent in Uttar Pradesh (Northern region) to–1.45 per cent in Maharashtra (western region). The cotton crop performed better in the western than southern region.

As in the case of other crops, the growth performance of groundnut and red gram was quite dismal, with declining production of former and near stagnation in the latter.

Table 2.9: Compound growth rates of area, production and yield of important crops across regions for the period from 1990-91 to 2005-06

States/Region	Rice			Sugarcane			Cotton			Groundnul			Redgram		
	Area	Prodn.	Yield	Area	Frodn.	Yield	Area	Prodn.	Yield	Area	Prodn.	Yield	Area	Prodn.	Yield
Southern															
Andhra Pradesh	-0.94	0.66	1.6	1.44	1.71	0.27	2.51	-1.58	-3.98	-2.92	-4.72	-1.87	3.47	8.27	4.58
Karnataka	0.46	1.95	1.49	-0.76	-0.97	-0.27	-2.48	-5.08	-2.06	-2.64	-4.16	-1.58	2.28	5.77	3.31
Kerala	-4.6	-3.92	1.15												
Tamil Nadu	-1.4	-2.78	-1.39	0.68	0.45	-0.17	-6.1	-9.16	-3.12	-4.85	-3.55	1.42	-7.34	-6.7	0.21
Western															
Maharashtra	-0.19	0.41	0.39	0.93	-0.53	-1.45	0.69	5.18	4.47				0.37	3.46	3.09
Gujarat	0.92	1.78	0.86	3.7	2.48	-1.06	4.26	5.22	0.92	0.34	5.86	5.49	-3.06	-2.75	0.16
Rajasthan							-1.21	-5.04	-3.87	0.52	3.5	3.35			
Madhya Pradesh	-9.48	-11.67	-2.42				0.77	-7.27	-7.98	-2.5	0.75	1.33	-2.27	-3.53	-1.38
Northern															
Uttar Pradesh	3.34	0.94	0.72	0.83	0.83	-0.002				-3.16	-3.98	-0.96	-2.63	-2.89	-0.29
Uttarakhand															
Punjab	1.99	3.01	1.01	0.35	0.23	-0.23	-2.55	-1.08	1.51						
Haryana	3.43	3.68	0.25	0.49	1.36	0.89	0.94	0.91	0.37						
Eastern															
Bihar and Jharkand	-3.31	-2.56	0.77	-2.7	-3.29	-0.88							-5.26	-4.63	0.93
West Bengal	0.07	2.1	2.04												
Orissa	-0.05	-0.07	-0.03										-2	-3.08	-1.24
North-eastern															
Assam	-0.07	1.09	1.16												
All-India	0.1	1.07	0.97	0.92	0.73	-0.21	0.78	0.39	-0.38	-2.21	-1.28	1.16	-0.12	0.3	0.41

Variations in Farm Profitability

The crop production, productivity and efficiency of farms varied significantly with managerial factors (Selvarajan *et al.*, 1997), farm size (Sen and Bhatia, 2004; Haque, 1996), farmers' access to technology, credit, market infrastructure, and price support (Haque, 2006). This, in turn, had a bearing on farm profitability wherein wide intra and inter regional differences have been observed (Singh, 2006).

In this Chapter, the profitability has been defined in terms of net income and the ratio of gross value of output to the cost of cultivation. The net return per hectare at constant prices (1993-94) was considered as an indicator of farm profitability The triennium averages of profitability for the years 2000-01 to 2002-03 are discussed below.

Rice

Rice is a major crop produced in almost all regions of the country with varying level of productivity and production. The per hectare cost of cultivation (Table 2.10), based on farmers' paid out costs (Cost A_2 + FL), ranged from Rs. 5207 in Madhya Pradesh (western region) to as high as Rs. 13329 in Tamil Nadu (southern region).

The differences in cost of cultivation have been attributed to the differences in levels of use of various inputs such as seeds, fertilizers, manure, human and animal labour, etc. Also, variations in the prices of these inputs matter in cost of cultivation and production. For instance, the use of seeds ranged from 63.22 kg/ha in Bihar (eastern region) to 98.4 kg/ha in Madhya Pradesh (western region) in 2000-01. Similarly, the fertilizer–use visualized inter-regional differences. The fertilizer-use/ha varied from very low level of 12.03 kg/ha in Assam, to moderate level of 50.71 kg/ha in Madhya Pradesh and further to as high as 222.16 kg/ha of nutrients in Karnataka (southern region) in 2000-01. The variations in input-use coupled with inter-regional variations in prices of critical inputs were the reasons for differences in cost of cultivation across regions.

The gross income ranged from Rs. 4977/ha in Madhya Pradesh (western region) to Rs. 26602/ha in Punjab (northern region). The net income showed a miserable picture. The per hectare profit has been found to be very low in many states and regions and even negative in Orissa, West Bengal (eastern region) and Madhya Pradesh (western region). Punjab (northern region) had advantage in

Table 2.10: Profitability of Rice Production Across Regions/States in India

(in Rs/ha)

Regions	*States*	*Yield (q/ha)*	*Cost C_2*	*Cost A_2+FL*	*Gross value of output*	*Net income over C_2*	*Net income over A_2+FL*	*GVCR**
Southern	Andhra Pradesh	48.46	17035	11258	16584	-450	5326	1.47
	Karnataka	43.80	17258	12658	15944	-1314	3286	1.26
	Kerala	33.91	15448	12153	13119	-2328	965	1.08
	Tamil Nadu	46.51	18216	13329	14724	-3491	1394	1.10
Western	Madhya Pradesh	15.27	7248	5207	4977	-2270	-230	0.96
Northern	Haryana	39.91	15438	10236	16565	1127	6329	1.62
	Punjab	58.55	15500	9787	26202	10701	16414	2.68
	Uttar Pradesh	32.28	9921	6926	8610	-1311	1683	1.24
Eastern	Bihar	23.62	7802	5267	5920	-1882	652	1.12
	Orissa	27.84	9982	7260	6886	-3096	-374	0.95
	West Bengal	33.77	13303	10006	9275	-4028	-730	0.93
North-eastern	Assam	25.46	8163	6034	8066	-96	2032	1.34
Mean		35.78	12943	9177	12239	-703	3062	1.31

*Indicates the ratio of gross value of output to the cost of cultivation (A_2+FL)

Sources: *Agricultural Statistics at a Glance* (2006), Directorate of Economics and Statistics, Ministry of Agriculture, Government of India, New Delhi.

Report of the Commission for Agricultural Costs and Prices (2004), Department of Agriculture and Co-operation, Ministry of Agriculture, Government of India, New Delhi.

producing rice, as evidenced from high net income of Rs. 16114/ha and the ratio of gross value of output to farmers paid out cost of 2.68 in terms of profitability, followed by Haryana (northern region) and Andhra Pradesh (southern region).The diversity of resources, agro-climatic conditions, and market infrastructure had provided profitability advantage to these few regions in producing rice.

Sugarcane

Being a commercial crop, sugarcane cultivation involves a number of operations, which escalate the cost of cultivation and production. The per hectare cost of cultivation of sugarcane varied from Rs. 10954 in Uttar Pradesh to Rs. 26965 in Tamil Nadu (Table 2.11). The high cost of cultivation in Tamil Nadu was mainly due to lbour scarcity for the sugarcane harvesting. The farmers in Tamil Nadu had to incur huge cost on sugarcane harvesting, which ranged from Rs. 230 to Rs. 300 per tonne of cane. Farm mechanization would be a viable solution to solve the labour problem in sugarcane cultivation. The cost of human labour alone worked out to be Rs. 25992/ha in Tamil Nadu, whereas it was Rs. 18111/ha in Andhra Pradesh, Rs. 11700/ha in Haryana, Rs. 13288/ha in Maharashtra, and Rs. 8429/ha in Uttar Pradesh. The variations in labour availability, and wage rate had affected the cost of cultivation. In addition, the level of input-use and their prices had also contributed to the differences in cost of cultivation across states and regions.

Differences in level of input-use had also resulted in higher productivity. Although the cost of cultivation was high in Tamil Nadu, it ranked first in productivity with a higher level of 1150 q/ha. The use of fertilizer nutrients showed that it was as high as 506.48 kg/ha in Tamil Nadu and as low as 134.30 kg/ha in Haryana (northern). The other northern state, viz. Uttar Pradesh had low level of fertilizer use (173.73 kg/ha) in 2000-01.

The gross income varied from Rs. 24071/ha in Uttar Pradesh to Rs. 56375/ha in Tamil Nadu. Thus, considerable differences could be visualized in gross income across states and regions. Similarly, the net income per hectare varied from Rs. 4594 in Maharashtra (western region) to Rs. 29410 in Tamil Nadu (southern region). The differences in both, cost of cultivation and prices of sugarcane led to the differences in profitability of sugarcane.

Table 2.11: Profitability of Sugarcane Production Across Regions/States in Idia

(in Rs/ha)

Regions	States	Yield (q/ha)	Cost C_2	Cost A_2+FL	Gross value of output	Net income over C_2	Net income verA$_2$+FL	GVCR*
Southern	Andhra Pradesh	727.30	32747	20685	37251	4504	16566	1.80
	Tamil Nadu	1150.26	40602	26965	56375	15772	29410	2.09
Western	Maharshtra	775.36	30914	22757	27352	-3562	4594	1.20
Northern	Haryana	589.35	27952	12354	38001	10049	25646	3.08
	Uttar Pradesh	451.08	18204	10954	24071	5866	13116	2.20
Mean		738.67	30084	18743	36610	6526	17866	2.07

*Indicates the ratio of gross value of output to the cost of cultivation (A_2+FL)

Source: Agricultural Statistics at a Glance (2006), Directorate of Economics and Statistics, Ministry of Agriculture, Government of India, New Delhi.

Report of the Commission for Agricultural Costs and Prices (2004), Department of Agriculture and Co-operation, Ministry of Agriculture, Government of India, New Delhi.

Cotton

The profitability analysis of cotton revealed that though cotton is considered a commercial crop, it was found to be non-remunerative to the farmers in many of the states and regions. The net returns over the farmers paid out costs worked out to be-Rs. 963/ha in Tamil Nadu, while these were fairly reasonable, Rs. 7677/ha in another southern state, Andhra Pradesh. The states like Gujarat (-Rs. 560/ha), Madhya Pradesh (-Rs. 534/ha) and Punjab (-Rs. 114/ha) had negative returns from cotton cultivation (Table 2.12).

The differences in profits were mainly due to variations in yield and prices of output in addition to the cost of cultivation. The price received by the farmers varied considerably, from Rs. 1586/q in Maharashtra to Rs. 2055/q in Karnataka during 2002-03 (current prices).

Being prone to pest and diseases, cotton cultivation involves use of higher amount of plant protection chemicals, which escalates the cost of its cultivation. Also, being a labour-intensive crop, a considerable proportion is spent on labour cost. Of the total operational cost, expenditure on plant protection chemicals accounted for 1.3 per cent in Madhya Pradesh (western region) and 32 per cent in Punjab (northern region). The expenditure on human labour varied considerably and ranged from 36 per cent in Punjab to 65 per cent in Tamil Nadu. Increasing demand for human labour coupled with high wage rate in the Tamil Nadu state were the reasons for high cost of cultivation in the state and consequent negative net returns.

Groundnut

Groundnut is an important crop and grown widely in most of the southern states and parts of western region covering Gujarat and Maharashtra. The profit earned per hectare of groundnut cultivation exhibited not only the inter-regional differences but also intra-regional differences. In the western region, the net income over the paid out cost ranged from-Rs. 1390/ha in Maharashtra to Rs. 1917/ha in Gujarat, primarily due to vast differences in cost of cultivation in the two states, as the yield of groundnut was not much different in these states (Table 2.13).

In the southern region, the net returns were about Rs. 1000/ha in Tamil Nadu, Rs. 600/ha in AP and near negligible in Karnataka.

Table 2.12: Profitability of Cotton Production Across Regions

(in Rs/ha)

Regions	*States*	*Yield (q/ha)*	*Cost C_2*	*Cost A_2+FL*	*Gross value of output*	*Net income over C_2*	*Net income overA_2+FL*	*GVCR**
Southern	Andhra Pradesh	14.49	15078	9528	17205	2126	7677	1.81
	Karnataka	5.15	7072	4926	6471	-600	1545	1.31
	Tamil Nadu	11.35	18798	12678	11714	-7084	-963	0.92
Western	Gujarat	6.24	8841	6922	6361	-2479	-560	0.92
	Madhya Pradesh	4.23	7848	5517	4983	-2865	-534	0.90
	Maharashtra	7.44	10770	8255	8460	-2310	204	1.02
Northern	Haryana	8.11	10932	7705	8410	-2521	705	1.09
	Punjab	10.57	15777	11501	11386	-4391	-114	0.99
Mean		8.45	11890	8379	9374	-2515	994	1.12
SD		3.47	4209	2732	3953	2666	2815	0.31
CV (%)		41.03	35.40	32.60	42.17	-106	283	27.59

* Indicates the ratio of gross value of output to the cost of cultivation (A_2+FL)

Source: *Agricultural Statistics at a Glance* (2006), Directorate of Economics and Statistics, Ministry of Agriculture, Government of India, New Delhi.

Report of the Commission for Agricultural Costs and Prices (2004), Department of Agriculture and Co-operation, Ministry of Agriculture, Government of India, New Delhi.

Table 2.13: Profitability of Groundnut Production Across Regions

(in Rs/ha)

Regions	*States*	*Yield (q/ha)*	*Cost C_2*	*Cost A_2+FL*	*Gross value of output*	*Net income over C_2*	*Net income over A_2+FL*	*GVCR**
Southern	Andhra Pradesh	8.62	8930	6481	7107	-1822	626	1.10
	Karnataka	6.26	6491	4855	4847	-1643	-7	1.00
	Tamil Nadu	14.70	15593	11094	12098	-3495	1003	1.09
Western	Gujarat	10.04	9036	6803	8720	-315	1917	1.28
	Maharashtra	11.11	12606	10153	8762	-3843	-1390	0.86
Mean		10.15	10531	7877	8307	-2224	429	1.07
SD		3.13	3572	2634	2653	1447	1233	0.15
CV (%)		30.82	33.92	33.45	31.94	-65.07	286.92	14.41

*Indicates the ratio of gross value of output to the cost of cultivation (A_2+FL)

Source: Agricultural Statistics at a Glance (2006), Directorate of Economics and Statistics, Ministry of Agriculture, Government of India, New Delhi.

Report of the Commission for Agricultural Costs and Prices (2004), Department of Agriculture and Co-operation, Ministry of Agriculture, Government of India, New Delhi.

Although the cost of cultivation was highest in Tamil Nadu, partly due to higher productivity as compared to other two southern states and partly because of better prices in comparison to Karnataka, the net returns were also highest in Tamil Nadu.

Red Gram

In red gram cultivation, the net income ranged from Rs. 1016/ha in Karnataka (southern region) to Rs. 6588 in Maharashtra (western region). The gross income worked out to be Rs. 4872/ha in Karnataka and Rs10979/ha in Maharashtra (Table 2.14). Adequate price support, varietal improvement, and technological support including inputs could help maximize farm profits and reduce inter-regional differences in profitability in red gram.

Table 2.14: Profitability of Redgram Production Across Regions

(in Rs/ha)

Regions	*States*	*Yield (q/ha)*	*Cost C_2*	*Cost A_2+FL*	*Gross value of output*	*Net income over C_2*	*Net income over A_2+FL*	*GVCR**
Southern	Andhra Pradesh	7.33	7980	5012	6449	-1531	1437	1.29
	Karnataka	5.14	5464	3855	4872	-591	1016	1.26
Western	Gujarat	8.07	7248	4755	7878	630	3123	1.66
	Madhya Pradesh	6.13	5446	3506	5207	-238	1700	1.48
	Maharashtra	8.83	6700	4390	10979	4278	6588	2.50
Northern	Uttar Pradesh	9.14	7040	3364	8377	1336	5013	2.49
Mean		7.44	6646	4147	7294	647	3146	1.78
SD		1.56	1013	676	2281	2036	2230	0.57
CV (%)		21.03	15.25	16.30	31.28	314	70	32.15

*Indicates the ratio of gross value of output to the cost of cultivation (A_2+FL)

Source: Agricultural Statistics at a Glance (2006), Directorate of Economics and Statistics, Ministry of Agriculture, Government of India, New Delhi.

Report of the Commission for Agricultural Costs and Prices (2004), Department of Agriculture and Co-operation,

Ministry of Agriculture, Government of India, New Delhi.

Farm Profitability–Agrarian Distress: A Nexus

Though it is hard to define agrarian distress in a single concept, this study has considered two indicators which may fairly reflect the agrarian distress, viz. number of farmers indebted and the percentage of rural population living below poverty line. The preliminary evidence of a nexus between farm profitability and agrarian distress was examined using the indebtedness and poverty estimates.

The availability of credit critically affects the production, productivity and farm profitability, and therefore a close look at the indebtedness of farmers across regions was essential. There was a wide regional variation in the percentage of farmers indebted and the average amount outstanding per indebted farmer (Table 2.15). On an average, 49 per cent of farmers were found indebted in the country, the percentage being highest in the southern region (72.70 per cent) and lowest in the north-east region (19.89 per cent). The all-India average of the amount outstanding per farmer was Rs. 12585 and it varied from Rs. 1447 in north-eastern region to Rs. 24992 in southern region. In the western and northern regions also the outstanding amount against the farmers was high. On the face of it, the higher incidence of indebtedness together with higher outstanding amount in the agriculturally better-off regions (e.g. northern, southern and western), while lower incidence and lower outstanding amount in the agriculturally backward regions (eastern and north-eastern) looks contradictory to the popular belief that agrarian distress and agricultural development tend to be inversely related. However, as the outstanding amount was not essentially indicative of unpaid overdues but signified the borrowings of the farmers, the incidence of indebtedness and outstanding amount captured the credit inflow in agriculture, which was obviously higher in agriculturally developed regions. A more detailed analysis, which was outside the scope of this study, is required to bring out the nexus between rural indebtedness and farm profitability.

The regional pattern of the incidence of rural poverty seems to confirm the link between farm profitability and agrarian distress. The percentage of rural population living below the poverty line was lower than the national average of 21.8 per cent in the agriculturally well-developed regions like southern and northern regions, while it was high in eastern India, where farm profitability was low. Besides profitability, production and productivity of crops, agrarian distress was determined by several other socio-economic, institutional and environmental factors which have been brought out by several researchers.

Table 2.15: Indicators of Agrarian Distress Across Regions

Regions	*States*	*Indebtedness*		*Poverty*
		Percentage of farm households indebted[a]	*Average amount outstanding per indebted farmer (Rs)*[a]	*Percentage of rural population below poverty line in 2004-05*[b]
Southern	Andhra Pradesh	82.0	23965	7.5
	Karnataka	61.6	18135	12.0
	Kerala	64.4	33907	9.6
	Tamil Nadu	74.5	23963	16.9
Western	Maharashtra	54.8	16973	22.2
	Gujarat	51.9	15526	13.9
	Rajasthan	52.4	18372	14.3
	Madhya Pradesh	50.8	14218	29.8
	Chattisgarh	40.2	4122	31.2
Northern	Uttar Pradesh	40.3	7425	25.3
	Punjab	65.4	41576	5.9
	Haryana	53.1	26007	9.2
	Jammu and Kashmir	31.8	1903	2.7
	Himachal Pradesh	33.4	9618	7.2
Eastern	Bihar	33	4476	32.9
	Jharkand	20.9	2205	40.2
	West Bengal	50.1	5237	24.2
	Orissa	47.8	5871	39.8
North-eastern	Assam	18.1	813	17.0
	Arunachal Pradesh	5.9	493	17.0
	Mizoram	23.6	1876	17.0
	Nagaland	36.5	1030	17.0
	Meghalaya	4.1	72	17.0
	Manipur	24.8	2269	17.0
	Sikkim	38.8	2053	17.0
	Tripura	49.2	2977	17.0
All-India		48.59	12585	21.8

Source: NSSO Report No. 498 (59/33/1) (2003) *Situation Assessment Survey of Farmers: Indebtedness of Farmer Households*, National Sample Survey 59th Round (January–December)

Agricultural Statistics at a Glance (2006), Directorate of Economics and Statistics, Ministry of Agriculture, Government of India, New Delhi.

a. Sidhu and Gill (2006)
b. *Poverty Estimates for 2004-05* (2007) Government of India, Press Information Bureau, March.
c. Singh *et al.* (2003)

Conclusions

The study has observed considerable differences in farm profits across states and regions. The differences in yields of crops, prices of inputs and outputs, costs of cultivation and resource endowments have been found the major reasons for the differences in farm profits across regions. The analysis has shown that variation is much higher in profitability of crops than yield and cost of cultivation of crops in most cases. The study has shown that variations in prices of output make significant differences in the farm profitability. Northern region covering Haryana, Punjab and Uttar Pradesh have profitability advantages in the production of rice, sugarcane, cotton and red gram. The southern region covering Andhra Pradesh, Karnataka and Tamil Nadu, has relatively more advantages in terms of profitability in producing rice, sugarcane, cotton and groundnut. Parts of western region covering Gujarat and Maharashtra have been found having advantages in producing groundnut and red gram.

References

Dhawan, B.D. (2000) Economic impact of major water resources projects on farm economy of India. *Agricultural Situation in India*. December.

Haque, T. (1996) *Sustainability of Smallholder Agriculture*: Concept Publishing Company, New Delhi

Haque, T. (2006) Resource use efficiency in Indian agriculture, *Indian Journal of Agricultural Economics*, **61** (1): 65-76.

Mathur, Archana S., Surajit Das and Subhalakshmi Sircar (2006) Status of agriculture in India: Trends and prospects, *Economic and Political Weekly*, **41** (52): 5327-5336.

Sardana, Parveen K., Veena Manocha and A.C. Gangwar. (1997) Growth and variations in agricultural performance in Haryana, *Indian Journal of Agricultural Economics*, **52** (3): 389-393.

Ramasamy, C. and K.N. Selvaraj (2001) Pulses, oilseeds and coarse cereals: Why they are slow growth crops?, *Indian Journal of Agricultural Economics*, 57(3): 290-315.

Ramasamy, C. (2003) *Constraints to Growth in Indian Agriculture: Needed Technology, Resource Management and Trade Strategies*, Presidential Address, 63rd Annual Conference of Indian Society of Agricultural Economics, 19-21 December, Orissa.

Rao, V. M. and D. V. Gopalappa (2004) Agricultural growth and farmer distress, *Economic and Political Weekly*, **39**(52): 5591-5597.

Sawant, S.D. (1997) Performance of Indian agriculture with special reference to regional variations, *Indian Journal of Agricultural Economics*, 52(3): 353-373.

Selvarajan, S., P.L. Agarwal, S. Pandey, F.P. Lansigan and S.K. Bandapadhaya (1997) System approach for analysing tradeoff between income, risk, and water use in rice-wheat production in northern India, *Field Crops Research*, 51: 147-161.

Sen, Abhijit and M.S. Bhatia (2004) *Cost of Cultivation and Farm Income: State of the Indian Farmer, A Millennium Study*, Vol.14, Department of Agriculture and Cooperation, Government of India, New Delhi.

Sidhu, R. S. and A. S. Bhullar (2006) Patterns and determinants of agricultural growth in the two Punjabs, *Economic and Political Weekly*, 40(53): 5620-5627.

Sidhu, R.S. and Sucha Singh Gill (2005) Agricultural credit and indebtedness in India: Some issues, *Indian Journal of Agricultural Economics*, 61(1): 11-35.

Singh, I.J, K.N. Rai and J.C. Karwasra (1997) Regional variations in agricultural performance in India, *Indian Journal of Agricultural Economics*, 52(3): 374-386.

Singh, J.P. (2006) Changing agrarian relationships in rural India, *Indian Journal of Agricultural Economics*, 61(1): 36-64.

3

Functioning of Lease Market and Smallholders: Evidence and Options

H. R. Sharma

Introduction

Land defines the limit to agricultural growth under the existing technology to a large extent. Given the existing stock of land resource, the issue of access to this resource is critical in determining the welfare of the farming community in India. The incidence of poverty is typically correlated with lack of access to land. Land plays a dual role in rural India. Aside from its value as a productive factor, land ownership confers collateral in credit markets, security in the event of natural hazards or life contingencies and social status. Those who control land tend to exert a disproportionate influence over rural institutions, including labour and credit markets.

Land distribution structure in rural India has undergone many changes, under the continuing demographic pressures, technological changes and market transactions and also because of land reforms, beginning as far back as the 1950s. The ownership of landholding is highly skewed in nature, which can be corrected to some extent by an efficient system of tenancy. Also, an effective land lease policy is a way to reform agrarian structure and activate the land market.

In this backdrop, this Chapter gives a brief overview of the provisions and implementation of tenancy legislation is India, and examines the macro level changes in different aspects of tenancy relations across states.

Land Lease Policy

The broad features of the land lease policy which was followed in the post-independent India got evolved through the recommendations of the Congress Agrarian Reform Committee, Report of Panel on Land Reforms, and the successive Five-Year Plans. Consequent to the recommendations of different committees, following national guidelines were laid down to enact tenancy legislations:

- To provide security of tenure; fixation of fair rent, one-fourth to one-fifth; landowners to be allowed to resume land for self-cultivation up to a limited area; the surrender of tenancy rights with mutual consent;
- Disabled persons, defence personnel and such exemptions be allowed to lease-out land;
- The term personal cultivation should be clearly defined if landlords are allowed to evict tenants to resume land for self-cultivation;
- Tenancy records should be corrected and oral tenancies should be abolished and on non-resumable land, the landlord-tenant nexus to be ended and tenants in these areas to be brought into direct contact with state.

According to the provisions in tenancy legislations, major states can be classified into following broad categories:

First, Kerala and Jammu and Kashmir have legally banned leasing-out of agricultural land without any exception whatsoever.

Second, in Telangana region of Andhra Pradesh leasing out of land by large holders is prohibited. Smallholdings are allowed to lease out land for a period of five years. In the Andhra region, leasing is permitted but regulated.

Third, Karnataka, Himachal Pradesh, Bihar, Gujarat and Uttar Pradesh have legally prohibited leasing out of agricultural land, excepting by certain disabled categories like widows, minors, armed personnel, etc.

Fourth, Punjab, Haryana, Maharashtra and Assam have not banned leasing. However, while in Punjab and Haryana tenants acquire rights to purchase land after six years of continuous possession, and in Maharashtra they acquire such rights within one year of the commencement of tenancy. Leasing is also permitted in Tamil Nadu but the law stipulates that every contract should be in written form and in triplicate. A copy of the document is required to be deposited with the revenue officials. In Rajasthan, the landowners can lease out land for a non-renewable period of five years.

Fifth, in Orissa all future leases were prohibited. However, past leases continue after surrendering half of the leased-in land to the landlords or rayat. In Madhya Pradesh, past leases were abolished but the future leases were permitted.

Sixth, in scheduled tribe areas of Andhra Pradesh, Bihar, Orissa, Madhya Pradesh and Maharashtra transfer of tribal land to non-tribals even on lease basis, can be permitted by competent authority.

Magnitude of Tenancy

The actual implementation of tenancy legislations has, fallen far short of its provisions. At the national level, 125.86 lakh tenants have been cumulatively conferred ownership rights till March, 2006 on area amounting to 167.14 lakh acres.

At the state level, the changes in the magnitude of tenancy in terms of proportions of different types of holdings and operated area leased-in (Table 3.1) show the following broad features:

First, the proportion of entirely–owned and self-cultivated holdings (EOH) increased by varying degree practically in all major states, except Bihar where it remained nearly unchanged.

Second, in so far as mixed holdings (MH) were concerned, there was no clear pattern across states. While proportion of such holdings decreased in Andhra Pradesh, Haryana, Jammu and Kashmir, Karnataka, Orissa, Punjab, Rajasthan, Tamil Nadu, Uttar Pradesh and West Bengal, not much change was discerned in the remaining states, except Bihar where it increased significantly.

Third, the proportion of entirely leased-in holdings (ELIH) decreased in most of the states with the notable exceptions of Andhra Pradesh, Orissa and Punjab where these increased.

Fourth, incidence of tenancy in terms of proportion of operated area leased-in (LIA) decreased in most of the states. The notable exceptions to this broad trend were Gujarat, Kerala, Bihar and Orissa where proportion of operated area leased-in increased and Andhra Pradesh where it remained nearly unchanged.

The changes in the proportion of households not operating land, which comprised households owning but not operating and neither owning nor operating land, also provided some insights about the changing incidence of tenancy. Table 3.2 gives information on changes in all the three categories of households during 1990s. It may be seen from Table 3.2 that the proportion of households not operating land increased hugely practically in all the states. In fact, in some states, most notably, Andhra Pradesh, Bihar, Gujarat, Haryana, Karnataka, Kerala, Maharashtra, Tamil Nadu, Uttar Pradesh and West Bengal, the increase in the proportion of such households was truly mind boggling. It may also be seen that in as many as eleven states (Assam, Bihar, Gujarat, Jammu and Kashmir, Kerala, Madhya Pradesh, Orissa, Punjab, Rajasthan, Uttar Pradesh and West Bengal) increase in households not operating

Table 3.1: Incidence of Tenancy in Rural India: Major States: 1991-92 to 2002-03

(per cent)

State	*Year*	*EOH*	*MH*	*ELIH*	*LIA*
Andhra Pradesh	1991-92	79.69	13.32	2.43	9.57
	2002-03	85.10	7.85	5.18	9.97
Assam	1991-92	79.88	5.07	7.35	8.87
	2002-03	87.84	5.05	3.05	5.06
Bihar	1991-92	86.02	1.90	5.56	3.91
	2002-03	86.88	8.86	3.93	8.90
Gujarat	1991-92	92.83	2.36	2.35	3.34
	2002-03	94.29	3.16	1.81	5.08
Haryana	1991-92	78.96	12.41	4.83	33.74
	2002-03	88.15	9.95	0.95	14.38
Himachal Pradesh	1991-92	84.08	3.80	5.01	4.83
	2002-03	93.24	4.05	2.30	2.87
Jammu & Kashmir	1991-92	91.20	5.35	0.52	3.73
	2002-03	97.88	1.35	0.60	0.32
Karnataka	1991-92	80.35	6.14	3.73	7.43
	2002-03	94.33	3.01	2.20	3.68
Kerala	1991-92	90.99	2.54	3.21	2.88
	2002-03	93.96	3.35	1.24	4.18
Madhya Pradesh	1991-92	80.21	5.06	6.48	6.30
	2002-03	00.01	5.71	1.54	3.42
Maharashtra	1991-92	89.22	5.53	2.40	5.48
	2002-03	90.71	4.71	2.30	4.59
Orissa	1991-92	70.46	17.69	1.70	9.48
	2002-03	77.00	12.00	8.40	13.15
Punjab	1991-92	80.27	15.60	1.43	10.83
	2002-03	86.50	10.30	2.40	17.84
Rajasthan	1991-92	90.32	6.22	1.55	5.19
	2002-03	96.75	2.11	0.81	2.81
Tamil Nadu	1991-92	74.56	11.02	5.82	10.89
	2002-03	89.41	5.78	3.71	6.10
Uttar Pradesh	1991-92	80.84	14.47	2.64	10.49
	2002-03	87.51	9.42	2.58	9.60
West Bengal	1991-92	75.40	12.00	5.27	10.40
	2002-03	83.59	9.00	5.10	9.42
All-India	1991-92	81.98	8.87	3.85	8.28
	2002-03	88.60	6.99	3.06	6.60

Notes: EOH-Entirely-owned and self-cultivated holdings, MH =Mixed holdings ELIH = Entirely leased-in holdings, LIA = Operated area leased-in

Table 3.2: Changing Proportions of Households not Operating Land, Owning But not Operating and Neither Owning Nor Operating Land: Major States 1991-92 To 2002-03

(in per cent)

State	*Households not operating land*		*Owning but Not operating*		*Neither owning Nor operating*	
	1991-92	*2002-03*	*1991-92*	*2002-03*	*1991-92*	*2002-03*
Andhra Pradesh	37.49	57.06	31.84	43.93	5.65	13.13
Assam	14.01	18.08	7.89	14.97	6.12	3.11
Bihar	18.62	28.74	10.99	22.57	7.63	6.17
Gujarat	25.11	38.22	15.31	26.06	9.80	12.16
Haryana	19.05	32.34	16.18	23.10	2.87	9.24
Himachal Pradesh	12.22	20.90	7.83	7.14	4.39	13.76
Jammu & Kashmir	6.39	7.65	5.24	4.82	1.15	2.83
Karnataka	19.04	38.56	13.71	25.06	5.33	13.50
Kerala	6.55	37.28	3.70	34.10	2.85	3.18
Madhya Pradesh	21.88	27.26	11.46	16.58	10.42	10.68
Maharashtra	32.00	43.68	22.51	27.78	9.49	15.90
Orissa	27.20	33.69	21.97	26.75	5.23	6.94
Punjab	25.62	31.42	21.73	27.94	3.89	3.48
Rajasthan	12.54	17.74	9.98	12.57	2.56	5.17
Tamil Nadu	36.66	59.23	27.92	44.13	8.74	15.10
Uttar Pradesh	11.75	17.42	9.27	14.13	2.48	3.29
West Bengal	15.31	32.04	8.68	27.86	6.63	4.18
All-India	21.84	33.97	15.68	25.30	6.16	8.67

Source: i. *Report on Some Aspects of Ownership Holdings*; 48th Round 1991-92, Report No. 399.

ii. *Report on Livestock and Implements in Household Operational Holdings*; 48th Round 1991-92, Report No.408

iii. *Report on Some Aspects of Operational Holdings, 2002-03*, 59th Round, NSS Report No. 492

land was largely contributed by those who owned but did not operate land. It also needs to be mentioned here that households who own but do not operate land is an amalgam of those who have leased-out their land, are not owning enough cultivable land and those who have kept their land fallow. In the remaining six states (Andhra Pradesh, Haryana, Himachal Pradesh, Karnataka, Maharashtra and Tamil Nadu), there was also a significant increase in the proportion of households who neither owned nor operated any land.

Farm-size and Tenancy Status: As far as leasing-in of land by holdings of different farm-size categories was concerned, in 2002-03, the correlation coefficient between farm-size and proportion of operational holdings leased-in (Table 3.3) was positive in ten major states, namely Andhra Pradesh, Assam, Gujarat, Haryana, Karnataka, Kerala, Punjab, Rajasthan, Tamil Nadu and Uttar Pradesh compared to 1991-92 when these coefficients were positive in seven states (Andhra Pradesh, Assam, Haryana, Jammu and Kashmir, Punjab, Uttar Pradesh and west Bengal).

Table 3.3: Correlation Between Farm-Size and Proportion of Leased-in Holdings and Farm Size and Proportion of Leased-in Area: Major States: 1991-92 To 2002-03

State	*1991-92*		*2002-03*	
	Farm size and proportion of operational holdings leased-in	Farm size and proportion of operated area leased-in	Farm size and proportion of operational holdings leased-in	Farm size and proportion of area leased-in
Andhra Pradesh	0.4033	0.3687	0.5291	0.4736
Assam	0.8648	0.8351	0.0070	-0.1167
Bihar	-0.5559	-0.8664	-0.5531	-0.5320
Gujarat	-0.1236	-0.3117	0.0142	-0.0437
Haryana	0.9902	0.9803	0.9039	0.5505
Himachal Pradesh	-0.7944	-0.8253	-0.3649	-0.4423
Jammu & Kashmir	0.0108	-0.1684	-0.2446	-0.2196
Karnataka	-0.2630	0.4960	0.0334	-0.1066
Kerala	0.0428	0.8553	0.4071	0.4498
Madhya Pradesh	-0.3602	-0.5163	-0.3697	-0.4129
Maharashtra	-0.2951	0.9515	-0.6063	-0.4391
Orissa	-0.4788	-0.7873	-0.5949	-0.5893
Punjab	0.0697	0.8933	0.3081	0.2979
Rajasthan	-0.3269	0.4627	0.8971	0.7179
Tamil Nadu	-0.3848	-0.6525	0.8455	0.8596
Uttar Pradesh	0.3761	-0.8103	0.7691	0.7087
West Bengal	0.0646	-0.4786	-0.3185	-0.2626
All-India	-0.4427	0.6205	0.7520	0.3053

The correlation coefficient between farm-size and proportion of operated area leased-in during 2002-03 was, however, positive in seven states (Andhra Pradesh, Haryana, Kerala, Punjab, Rajasthan, Tamil Nadu and Uttar Pradesh) compared to 1991-92 when it was positive in ten states (Andhra Pradesh, Assam, Haryana, Karnataka, Kerala, Maharashtra, Punjab and Rajasthan).

The marginal and small holdings accounted for more than 75 per cent of the leased-in land (Table 3.4) in seven states (Bihar, Himachal Pradesh, Jammu and Kashmir, Kerala, Orissa, Tamil Nadu and West Bengal). In two states (Assam and Uttar Pradesh), the per cent share of such holdings varied from 50 per cent to 75 per cent, while in four others (Andhra Pradesh, Karnataka, Madhya Pradesh and Maharashtra), it was between 25 per cent and 50 per cent. In the remaining states (Gujarat, Haryana, Rajasthan and Punjab), it was less than 25 per cent.

Table 3.4: Classifications of States According to Share of Marginal and Small Holdings in the Total Area Leased-in Major States: 1991-92 To 2002-03

Share	*1991-92*	*2002-03*
> 75 per cent	Bihar, Himachal Pradesh, West Bengal	Bihar, Himachal Pradesh, Jammu & Kashmir, Kerala, Orissa, Tamil Nadu, West Bengal
50-75 per cent	Jammu & Kashmir, Kerala, Orissa, Tamil Nadu and Uttar Pradesh	Assam and Uttar Pradesh
25 to 50 per cent	Andhra Pradesh, Assam and Madhya Pradesh	Andhra Pradesh, Karnataka, Madhya Pradesh and Maharashtra
< 25 per cent	Gujarat, Haryana, Maharashtra, Punjab and Rajasthan	Gujarat, Rajasthan, Punjab and Haryana

Terms of Tenancy

The relative importance of different terms of tenancy can be understood through changes in the per cent distribution of leased-in holdings and operated area leased-in under different terms of tenancy (Table 3.5).

The proportion of holdings and area leased-in accounted for by the fixed rent tenancy increased in as many as 13 states (Andhra Pradesh, Bihar, Gujarat, Haryana, Himachal Pradesh, Karnataka,

Table 3.5: Distribution of Operational Holdings Leased-in and Operated Area Leased-in Under Different Terms of Tenancy in Rural India, Major States: 1991-92 To 2002-03

(in per Cent)

State	*Year*	*Holdings*				*Area*			
		FM	*FP*	*ST*	*OT*	*FM*	*FP*	*ST*	*OT*
Andhra	1991-92	26.70	27.65	29.81	15.84	25.90	26.80	28.90	18.40
Pradesh	2002-03	30.68	37.23	16.80	15.29	27.88	33.57	33.51	5.04
Assam	1991-92	19.40	4.57	31.17	44.86	17.00	4.00	27.80	51.20
	2002-03	9.92	3.09	58.20	28.79	10.73	2.26	33.33	53.68
Bihar	1991-92	10.02	13.35	46.05	30.58	9.50	12.80	43.50	34.20
	2002-03	16.67	20.72	57.15	5.46	12.85	18.52	64.85	3.78
Gujarat	1991-92	44.46	1.79	26.52	27.23	39.90	1.60	23.70	34.80
	2002-03	45.73	6.53	31.66	16.08	9.40	47.65	36.68	6.27
Haryana	1991-92	67.26	5.70	21.17	5.33	61.40	5.20	19.90	13.50
	2002-03	69.48	9.98	13.89	6.65	73.52	8.61	15.00	2.87
Himachal	1991-92	14.99	9.91	11.94	63.16	13.64	8.98	10.87	66.51
Pradesh	2002-03	36.88	15.60	30.50	17.02	20.00	13.33	26.67	40.00
Jammu &	1991-92	9.75	31.05	12.27	46.93	9.37	30.21	11.81	48.61
Kashmir	2002-03	40.00	0.00	20.00	40.00	7.62	6.09	64.55	21.74
Karnataka	1991-92	23.15	16.58	32.42	27.85	20.40	14.70	28.60	36.30
	2002-03	36.89	36.58	22.65	3.88	33.53	41.18	23.53	1.76
Kerala	1991-92	19.02	-	2.48	78.50	15.90	-	2.10	82.00
	2002-03	37.72	10.18	10.78	41.32	40.39	7.57	15.56	36.48
Madhya	1991-92	17.21	24.17	28.05	30.57	15.3	21.40	24.90	38.4
Pradesh	2002-03	24.79	36.34	29.86	9.01	19.85	30.64	36.76	12.75
Maharashtra	1991-92	39.59	7.07	22.94	30.40	36.20	6.50	20.90	36.40
	2002-03	31.44	8.68	40.12	19.76	28.72	9.22	40.04	22.01
Orissa	1991-92	20.58	5.83	53.29	20.30	19.70	4.70	50.90	24.70
	2002-03	11.88	9.98	68.52	9.62	10.97	9.19	71.61	7.81
Punjab	1991-92	52.56	19.50	12.07	15.87	49.20	18.20	11.30	21.30
	2002-03	81.73	3.25	8.51	6.50	80.43	1.52	14.43	3.62
Rajasthan	1991-92	15.78	20.18	24.31	39.73	15.2	19.40	23.40	42.00
	2002-03	34.82	12.50	41.96	10.71	36.94	18.47	36.94	7.65
Tamil Nadu	1991-92	33.52	21.13	16.67	28.68	32.40	20.50	16.10	31.00
	2002-03	42.67	28.22	15.00	14.11	32.68	29.53	22.44	15.35
Uttar	1991-92	9.80	16.21	49.48	24.45	9.20	15.20	46.50	29.10
Pradesh	2002-03	20.37	14.37	59.85	5.41	24.25	14.12	54.61	7.02
West	1991-92	9.61	13.02	51.77	25.60	8.60	11.70	46.50	33.20
Bengal	2002-03	26.71	29.96	35.39	7.94	24.92	29.26	33.50	12.23
All-India	1991-92	20.32	15.62	36.89	27.17	19.00	14.50	34.40	32.10
	2002-03	26.57	19.80	43.96	9.67	30.19	19.88	40.60	9.33

Source: i. *Report on Some Aspects of Operational Holdings*; 1991-92 48th Round, Report No. 407

ii. *Report on Some Aspects of Operational Holdings*, 2002-03, 59th Round, NSSO Report No. 492

Notes: FM: Fixed money, FP: Fixed produce, ST: Share tenenacy, OT: Others

Kerala, Madhya Pradesh, Punjab, Rajasthan, Tamil Nadu, Uttar Pradesh and West Bengal). Among the remaining states, while the proportion of holdings and area leased-in accounted for by fixed rent tenancy decreased in Assam and Jammu and Kashmir, no neat pattern was discernible in Maharashtra and Orissa.

The proportion of holdings and operated area leased-in accounted for by the share tenancy increased in eleven states (Assam, Bihar, Gujarat, Himachal Pradesh, Jammu and Kashmir, Kerala, Madhya Pradesh, Maharashtra, Orissa, Rajasthan and Uttar Pradesh) and decreased in three states, viz. Haryana, Karnataka and West Bengal. In other two states, namely, Punjab and Tamil Nadu while the proportion of holdings accounted for by share tenancy decreased, the proportion of operated area leased-in accounted for by it increased.

Farm-size and Terms of Tenancy: It is quite often argued in the literature that marginal and small holdings lease-in most of the land under share tenancy which is considered inferior contractual arrangement compared to fixed rent tenancy. To look into some of these issues, we computed the correlation coefficients between farm size and proportion of operated area leased-in under fixed money, fixed produce and share of produce (Table 3.6).

Table 3.6: Correlation Between Farm Size and Proportion of Operated Area Leased-in Under Different Terms of Tenancy: Major States: 1991-92 To 2002-03

State	*1991-92*			*2003*		
	Fixed money	*Fixed produce*	*Share of produce*	*Fixed money*	*Fixed produce*	*Share of produce*
Andhra Pradesh	-0.2248	-0.4619	0.6588	-0.3161	-0.3867	0.3841
Assam	0.8711	-0.1954	0.0057	0.1729	-0.3130	-0.2555
Bihar	-0.6575	-0.7177	-0.7965	-0.4140	-0.4490	-0.3968
Gujarat	0.0861	0.0725	0.1879	-0.3042	0.4841	0.3360
Haryana	0.2526	-0.3185	-0.0991	0.6755	0.1530	-0.2325
Himachal Pradesh	-0.4574	-0.4918	-0.4117	-0.4493	-0.3051	-0.2398
Jammu & Kashmir	-0.1709	0.0248	-0.1970	-0.1622	-0.1056	-0.2123
Karnataka	-0.1902	-0.0744	-0.0110	-0.4665	0.1446	-0.3455
Kerala	-0.4342	0.00	0.2821	0.4730	-0.3899	-0.3237
Madhya Pradesh	0.6152	0.2670	-0.0469	0.3575	-0.4496	-0.3569
Maharashtra	-0.8449	-0.3272	0.7953	-0.2907	0.0378	-0.0714
Orissa	-0.6685	-0.4670	-0.8401	-0.2997	-0.3863	-0.4038
Punjab	0.8000	-0.3902	-0.342	0.3034	-0.0858	0.1748
Rajasthan	-0.0512	0.9431	0.1188	0.5189	0.6137	-0.0903
Tamil Nadu	0.2004	-0.6793	0.1417	-0.4325	-0.3163	0.9004
Uttar Pradesh	-0.6433	0.0635	-0.2702	-0.1023	-0.3965	-0.4332
West Bengal	0.9482	-0.5124	0.4770	-0.0821	-0.4339	-0.3649
All-India	0.8276	-0.2560	-0.5815	0.2090	0.3081	0.2761

In eight states (Bihar, Gujarat, Himachal Pradesh, Kerala, Maharashtra, Orissa, Tamil Nadu and West Bengal), the correlation coefficients between farm size and proportion of operated area leased-in under fixed money were negative, implying that lower farm-size categories leased-in higher proportion under fixed money compared to their large counterparts, while, these were positive in the remaining nine states (Andhra Pradesh, Assam, Haryana, Jammu and Kashmir, Madhya Pradesh, Punjab, Rajasthan and Uttar Pradesh). In as many as eleven states (Andhra Pradesh, Assam, Bihar, Himachal Pradesh, Kerala, Madhya Pradesh, Orissa, Punjab, Tamil Nadu, Uttar Pradesh and West Bengal), the correlation coefficients between farm size and proportion of operated area leased-in under fixed produce were negative, while in other states these were positive.

The relationship between farm size and proportion of operated area leased-in under share produce was, however, negative in twelve states (Assam, Bihar, Haryana, Himachal Pradesh, Karnataka, Kerala, Madhya Pradesh, Maharashtra, Orissa, Rajasthan Uttar Pradesh and West Bengal), implying that lower farm-size categories leased-in higher proportion of land under share of produce compared to higher farm-size categories. Likewise, per cent share of marginal and small holdings in the land leased-in under fixed money, fixed produce and share of produce (Table 3.7), shows that in the states like Bihar, Kerala, Himachal Pradesh, Jammu and Kashmir, Tamil Nadu and West Bengal, more than 75 per cent of the land leased-in under fixed money was accounted for by these holdings. In five other states (Assam, Gujarat, Karnataka, Orissa and Uttar Pradesh), their share was between 50 per cent and 75 per cent. Likewise, in eight states (Assam, Bihar, Kerala, Himachal Pradesh, Jammu and Kashmir, Orissa, Tamil Nadu and West Bengal), these holdings accounted for more than 75 per cent of the land leased-in under fixed produce, followed by Andhra Pradesh and Madhya Pradesh, where it varied between 50 per cent and 75 per cent. In so far as their share in the area leased-in under share of produce was concerned, it was more than 75 per cent in Assam, Bihar, Himachal Pradesh, Kerala, Orissa, Uttar Pradesh and West Bengal. In Jammu and Kashmir and Madhya Pradesh, it was between 50 per cent and 75 per cent, whereas in Karnataka, Maharashtra and Tamil Nadu, they accounted for 25-50 per cent of the area leased-in under share of produce. In other states, it was less than 25 per cent.

Table 3.7: Classification of States According to Share of Marginal and Small Holdings in Land Leased-in under Different terms of Tenancy: Major States: 1991-92 To 2002-03

Share	*1991-92*	*2002-03*
	Fixed Money	
> 75 per cent	Bihar, Himachal Pradesh, Jammu & Kashmir, Kerala, Orissa and West Bengal	Bihar, Kerala, Himachal Pradesh, Jammu & Kashmir, Tamil Nadu and West Bengal
50-75 per cent	Uttar Pradesh	Assam, Gujarat, Karnataka, Orissa and Uttar Pradesh
25-50 per cent	Andhra Pradesh, Assam, Punjab and Tamil Nadu	Andhra Pradesh, Madhya Pradesh
< 25 per cent	Gujarat, Haryana, Karnataka, Maharashtra, Punjab and Rajasthan	Haryana, Punjab, Maharashtra and Rajasthan
	Fixed Produce	
> 75 per cent	Bihar, Himachal Pradesh, Jammu & Kashmir, Tamil Nadu and West Bengal	Assam, Bihar, Himachal Pradesh, Jammu & Kashmir, Kerala, Orissa, Tamil Nadu and West Bengal
50-75 per cent	Andhra Pradesh, Assam, Punjab and Uttar Pradesh	Andhra Pradesh and Uttar Pradesh
25-50 per cent	Haryana, Karnataka, Madhya Pradesh and Punjab	Andhra Pradesh, Karnataka, Madhya Pradesh and Maharashtra
<25 per cent	Gujarat, Maharashtra and Rajasthan	Gujarat, Haryana, Punjab and Rajasthan
	Share of Produce	
>75 per cent	Bihar and West Bengal	Assam, Bihar, Himachal Pradesh, Kerala, Orissa, Uttar Pradesh and West Bengal
50-75 per cent	Assam, Himachal Pradesh, Punjab, Orissa, Uttar Pradesh and Tamil Nadu	Jammu & Kashmir and Madhya Pradesh
25-50 per cent	Andhra Pradesh, Jammu & Kashmir, Karnataka and Kerala	Karnataka, Maharashtra and Tamil Nadu
< 25 per cent	Gujarat, Haryana, Maharashtra, Punjab and Maharashtra	Andhra Pradesh, Gujarat, Haryana, Punjab and Rajasthan

Reverse Tenancy

The extent of reverse tenancy has been studied in terms of changes in the amount of land leased-in and leased-out by the small and marginal households and their per cent share in the total land transacted in the lease market (Table 3.8).

Table 3.8: Reverse Tenancy: The Extent of Participation of Landless, Marginal and Small Households in the Lease Market as Lessees and Lessors: Major States: 1991-92 to 2002-03

State	Year	Land leased-in ('00 ha)	Land leased-out ('00 ha)	Share in leased-in land (%)	Share in leased-out land (%)
Andhra Pradesh	1991-92	6969	2396	82.52	64.17
	2002-03	6128	1607	71.74	70.57
Assam	1991-92	1804	402	95.04	58.69
	2002-03	1094	35	82.97	53.50
Bihar	1991-92	3118	981	95.03	41.50
	2002-03	5967	771	97.38	65.09
Gujarat	1991-92	2571	940	90.31	38.68
	2002-03	2426	833	70.74	61.93
Haryana	1991-92	4074	4234	28.76	64.06
	2002-03	1845	692	40.50	37.58
Himachal Pradesh	1991-92	-	-	97.57	65.15
	2002-03	194	126	94.78	91.58
Jammu & Kashmir	1991-92	166	-	96.51	70.71
	2002-03	23	1.52	73.08	93.55
Karnataka	1991-92	4187	1459	58.47	19.49
	2002-03	1588	743	66.53	22.53
Kerala	1991-92	262	-	82.13	-
	2002-03	428	57	98.74	63.97
Madhya Pradesh	1991-92	9084	1783	75.86	28.94
	2002-03	4350	1055	70.60	36.85
Maharashtra	1991-92	4734	1638	59.56	24.49
	2002-03	3301	1050	55.74	37.62
Orissa	1991-92	4484	1736	91.77	75.02
	2002-03	4560	1383	97.62	62.89
Punjab	1991-92	1767	810	42.41	32.36
	2002-03	2337	604	48.02	25.89
Rajasthan	1991-92	5147	709	57.37	10.99
	2002-03	1609	471	32.33	23.16
Tamil Nadu	1991-92	3900	1745	86.91	64.57
	2002-03	1819	773	66.94	76.15
Uttar Pradesh	1991-92	16767	4158	91.37	49.30
	2002-03	11890	4583	81.24	58.12
West Bengal	1991-92	5029	1145	97.20	60.45
	2002-03	3517	643	94.40	63.97
All-India	1991-92	74772	19497	71.27	32.44
	2002-03	53638	15637	71.00	47.73

Source: (i) *Report on Some Aspects of Household Ownership Holdings* (1):1991-92: 48th Round, NSSO Report No. 399
(ii) *Report on Household Ownership Holdings in India*: 2003, 59th Round NSSO Report No. 491

In as many as ten states (Andhra Pradesh, Assam, Gujarat, Jammu and Kashmir, Madhya Pradesh, Maharashtra, Rajasthan, Tamil

Nadu, Uttar Pradesh and West Bengal), the proportion of leased-in land accounted for by landless, marginal and small households decreased during 1991-92 to 2002-03. More importantly, however, in all these states, except Madhya Pradesh, Maharashtra and West Bengal, the extent of decrease in their per cent share in leased-in area was truly significant. In the remaining states, the proportion of leased-in land accounted for by these households increased.

At the all-India level, the leased-in land accounted for by landless, marginal and small households remained around 70 per cent. Insofar as share of small and marginal households in the total land supplied in the lease market was concerned, two distinct patterns were discernible. First, in as many as twelve states (Andhra Pradesh, Bihar, Gujarat, Jammu and Kashmir, Himachal Pradesh, Karnataka, Madhya Pradesh, Maharashtra, Rajasthan, Tamil Nadu, Uttar Pradesh and West Bengal), per cent contribution of small and marginal households towards total land supplied in the lease market increased during the 1990s. Second, in four other states (Assam, Haryana, Orissa and Punjab), the contribution of small and marginal households towards the total land leased-out land decreased. At the all-India level, there was a huge increase, from about 32 per cent to 48 per cent, in the proportion of leased-out land accounted for by small and marginal households.

The per cent share of marginal and small households in the total land leased-in and leased-out (Table 3.9) shows that in 2002-03 landless, marginal and small households accounted for more than 75 per cent of the land leased-in in seven states, namely Assam, Bihar, Himachal Pradesh, Kerala, Orissa, Uttar Pradesh, and West Bengal. In seven other states (Andhra Pradesh, Gujarat, Jammu and Kashmir, Karnataka, Madhya Pradesh, Maharashtra and Tamil Nadu), the share ranged between 50 per cent and 75 per cent of the leased-in land. In the remaining states (Haryana, Punjab and Rajasthan), it was 25-50 per cent. Insofar as the share of marginal and small households in the leased-out land was concerned, it was more than 75 per cent in three states (Himachal Pradesh, Jammu and Kashmir and Karnataka), followed by eight others (Andhra Pradesh, Assam, Bihar, Gujarat, Kerala, Orissa, Uttar Pradesh, and West Bengal), where it was between 50 per cent and 75 per cent. In four states (Haryana, Madhya Pradesh, Maharashtra and Punjab) these households supplied between 25 per cent and 50 per cent of the total leased-out land, while in two others (Karnataka and Rajasthan), the share was less than 25 per cent.

Table 3.9: Classification of States According to Share of Lower Category Households in Total Land Leased-in and Leased-Out: Major States: 1991-92 To 2002-03

Share (%)	*1991-92*	*2002-03*
Share of Landless, Marginal and Small Households in Total Land Leased-in		
> 75 per cent	Andhra Pradesh, Assam, Bihar, Gujarat, Himachal Pradesh, Jammu & Kashmir, Kerala, Madhya Pradesh, Orissa, Tamil Nadu, Uttar Pradesh, and West Bengal	Assam, Bihar, Himachal Pradesh, Kerala, Orissa, Uttar Pradesh and West Bengal
50-75 per cent	Karnataka, Maharashtra and Rajasthan	Andhra Pradesh, Gujarat, Jammu & Kashmir, Karnataka, Madhya Pradesh, Maharashtra, and Tamil Nadu
25-50 per cent	Haryana and Punjab	Haryana, Punjab and Rajasthan
< 25 per cent	–	–
Share of Marginal and Small Households in Total Land Leased-out		
> 75 per cent	Orissa	Himachal Pradesh, Jammu & Kashmir and Tamil Nadu
50-75 per cent	Andhra Pradesh, Assam, Haryana, Himachal Pradesh, Jammu & Kashmir, Tamil Nadu and West Bengal	Andhra Pradesh, Assam, Bihar, Gujarat, Kerala, Orissa, Uttar Pradesh and West Bengal
25-50 per cent	Bihar, Gujarat, Madhya Pradesh, Punjab and Uttar Pradesh	Haryana, Madhya Pradesh, Maharashtra and Punjab
< 25 per cent	Karnataka, Maharashtra and Rajasthan	Karnataka and Rajasthan

Duration of Tenancy Contracts

Duration of tenancy contracts has important implications towards agricultural productivity. As is well known, long term tenancy contracts, *ceteris peribus*, offer more incentives for undertaking productivity enhancement measures. Micro studies from different regions of the country report that most of the tenancy contracts are for a short period. The landowners, apprehending losing land to tenants, reportedly lease-out for one to two crop seasons and rotate tenants from one plot to others. The evidence from NSSO data (Table 3.10), lend credence to these findings of micro studies. The data has shown that in ten states (Assam, Bihar, Haryana, Karnataka, Madhya Pradesh, Maharashtra, Orissa, Punjab, Rajasthan and Uttar

Pradesh), more than half of the leased-in land was governed by contracts for less than two years. In the remaining seven states, namely, Andhra Pradesh, Gujarat, Himachal Pradesh, Jammu and Kashmir, Kerala, Tamil Nadu and West Bengal, more than half of the leased-in land was under contracts extending beyond two years. Further, during the 1990s, the duration of tenancy contracts, in terms of proportion of area leased-in under contracts of different durations, increased in five states (Andhra Pradesh, Gujarat, Haryana, Orissa and Tamil Nadu), but decreased in as many as twelve states, namely, Assam, Bihar, Haryana, Jammu and Kashmir, Karnataka, Kerala, Madhya Pradesh, Maharashtra, Punjab, Rajasthan, Uttar Pradesh and West Bengal.

Variations across Farm-size: Regarding relationship between farm size and the proportion of land leased-in under tenancy contracts of different durations (Table 3.11) shows that the correlation coefficients were negative in all the states. This suggests that lower size holdings leased-in higher proportion of land under contracts of different durations compared to large holdings.

Coming to the per cent share of marginal and small holdings in the land leased-in under contracts for different durations, Table 3.12 shows that in seven states (Himachal Pradesh, Karnataka, Kerala, Orissa, Tamil Nadu, Uttar Pradesh and West Bengal), more than 75 per cent of the leased-in area for less than two years was accounted for by such holdings, followed by states like Andhra Pradesh, Bihar, Madhya Pradesh, Gujarat and Jammu and Kashmir, where share of these holdings varied between 50 per cent and 75 per cent and Karnataka and Maharashtra where it was between 25 per cent and 50 per cent. In land leased-in for 2 to 5 years, these holdings accounted for 50 to 75 per cent of such land in Himachal Pradesh, Jammu and Kashmir, Tamil Nadu, Orissa and West Bengal, followed by Assam, Bihar, Kerala and Uttar Pradesh where their share varied between 25 per cent and 50 per cent. In the case of land leased-in for more than five years, in six states, the per cent share of such holdings was more than 75 per cent in six states (Andhra Pradesh, Jammu and Kashmir, Karnataka, Orissa, Tamil Nadu and West Bengal), followed by four states (Assam, Bihar, Himachal Pradesh and Kerala), where it varied from 50 per cent to 75 per cent and Karnataka, Maharashtra and Tamil Nadu, where it was between 25 per cent and 50 per cent.

Table 3.10: Distribution of Leased-in Area by Period of Lease: Major States: 1991-92 to 2002-03

(in per cent)

State	Year	Less than one agricultural season	One agricultural season but less than one agricultural year	One to less than two agricultural years	Two to less than five years	Five to less than twelve years	Twelve and more than twelve years	Not recorded	Total
Andhra Pradesh	1991-92	2.31	25.50	17.27	28.8	8.42	10.04	7.66	100
	2002-03	6.01	13.05	23.76	22.27	29.81	5.04	0.05	100
Assam	1991-92	0.84	29.05	15.26	14.49	1.93	6.33	32.01	100
	2002-03	5.97	19.79	23.24	15.93	9.70	5.44	19.93	100
Bihar	1991-92	3.64	9.30	34.26	17.82	7.76	7.55	19.40	100
	2002-03	2.13	16.51	45.78	19.78	9.27	5.09	1.44	100
Gujarat	1991-91	3.54	20.33	20.26	15.29	1.17	15.23	24.17	100
	2002-03	1.25	9.11	10.56	30.75	0.01	48.32	0.00	100
Haryana	1991-92	10.24	15.43	59.51	0.77	0.04	0.72	13.29	100
	2002-03	6.38	16.42	49.24	24.27	3.27	0.42	0.00	100
Himachal Pradesh	1991-92	1.18	5.60	6.47	12.25	13.14	48.32	13.03	100
	2002-03	2.02	10.71	8.38	22.22	24.28	32.38	0.00	100
Jammu & Kashmir	1991-92	7.69	0.00	9.16	13.86	4.98	48.58	15.72	100
	2002-03	3.31	25.32	16.17	50.00	0.24	4.98	0.00	100
Karnataka	1991-92	9.72	25.88	8.92	16.11	5.22	4.45	29.69	100
	2002-03	7.21	13.32	35.47	32.94	10.60	0.47	0.00	100
Kerala	1991-92	6.34	10.75	2.90	16.50	24.26	20.84	18.41	100
	2002-03	1.85	9.70	15.84	34.14	24.25	14.22	0.00	100

Contd.

Madhya Pradesh	*1991-92*	*0.95*	*14.81*	*25.09*	*8.77*	*7.03*	*20.56*	*22.78*	*100*
	2002-03	10.04	23.85	29.48	14.45	11.52	9.01	1.65	100
Maharashtra	1991-92	9.75	11.12	25.17	19.63	10.90	10.34	13.10	100
	2002-03	4.35	29.61	21.49	13.54	19.21	11.44	0.35	100
Orissa	1991-92	1.49	30.88	26.31	14.37	11.32	4.79	10.33	100
	2002-03	1.58	19.69	33.68	27.39	8.24	8.61	0.82	100
Punjab	1991-92	6.96	12.12	41.53	11.25	12.92	4.62	10.60	100
	2002-03	4.60	13.65	48.84	9.52	19.80	3.59	0.00	100
Rajasthan	1991-92	3.16	15.00	29.25	26.94	10.22	9.29	6.15	100
	2002-03	2.82	13.63	40.01	37.59	3.44	2.50	0.00	100
Tamil Nadu	1991-92	3.45	6.76	17.30	23.33	9.34	21.56	18.36	100
	2002-03	3.50	3.47	18.43	22.38	25.74	26.48	0.00	100
Uttar Pradesh	1991-92	2.64	15.08	38.83	11.81	10.48	9.43	11.80	100
	2002-03	3.52	23.65	36.76	17.57	7.55	7.66	3.29	100
West Bengal	1991-92	3.22	8.45	10.67	18.58	12.67	35.81	10.60	100
	2002-03	6.85	15.58	24.14	16.12	13.53	20.58	3.21	100
All-India	1991-92	4.82	16.44	30.67	14.64	7.81	10.87	14.72	100
	2002-03	4.62	17.96	32.91	20.67	12.82	9.52	1.49	100

Source: i. *Report on Some Aspects of Operational Holdings;* 1991-92, 48th Round Report No. 407

ii. *Report on Some Aspects of Operational Holdings, 2002-03,* 59th Round, NSS Report No. 492

Table 3.11: Relationship Between Farm Size and Proportion of Operated Area Leased-in Under Contracts of Different Duration (Correlation Coefficient): Major States: 1991-92 To 2002-03

State	*1991-92*			*2003*		
	Less than 2 years	*2 to 5 years*	*More than 5 years*	*Less than 2 years*	*2 to 5 years*	*More than 5 years*
Andhra Pradesh	-0.1691	0.3081	0.4570	-0.4344	-0.00	-0.2401
Assam	0.6664	-0.3679	-0.3995	-0.5384	-0.2919	-0.2586
Bihar	-0.3363	-0.5549	-0.5558	-0.2584	-0.2290	-0.2707
Gujarat	-0.2072	-0.7908	-0.8180	-0.2186	-0.0293	-0.1157
Haryana	-0.8786	-0.3623	-0.3623	-0.2264	-0.2487	-0.4658
Himachal Pradesh	-0.4112	-0.4616	-0.6111	-0.3906	-0.3066	-0.4169
Jammu & Kashmir	-0.3139	-0.4890	-0.4526	-0.3127	-0.2833	-0.2272
Karnataka	-0.5494	-0.3989	-0.4167	-0.1897	-0.0046	-0.3270
Kerala	-0.5835	-0.5449	-0.5428	-0.2766	-0.2713	-0.3749
Madhya Pradesh	-0.5975	-0.4127	-0.4869	-0.4380	-0.1563	-0.3194
Maharashtra	-0.5503	-0.3533	-0.5038	-0.0139	-0.2051	-0.5031
Orissa	-0.6067	-0.5064	-0.3942	-0.3882	-0.4687	-0.4170
Punjab	-0.5899	-0.2100	-0.3762	-0.1549	-0.1091	-0.2768
Rajasthan	-0.5595	-0.3120	-0.5482	-0.2444	-0.3795	-0.1304
Tamil Nadu	-0.5072	-0.5198	-0.5257	-0.4295	-0.2456	-0.2609
Uttar Pradesh	-0.5926	-0.5083	-0.5546	-0.5051	-0.3311	-0.2155
West Bengal	-0.5503	-0.5543	-0.6072	-0.3637	-0.3533	-0.4753
All-India	-0.5884	-0.5802	-0.5886	-0.1960	-0.2096	-0.2328

Micro Evidence

A number of micro studies have looked into various aspects of tenancy relations like magnitude, type, relative efficiency of different land tenure systems, and so on. Insofar as actual functioning of the lease market in rural areas is concerned, the evidence from such studies covering different states/regions of the country shows that the proportion of leased-in land is significantly higher than reported by both NSS and Census Data. In some cases, it is as high as 20-25 per cent of the gross cultivated area. Tenancy contracts are oral and for a short period. The proportion of leased-in land is higher in agriculturally developed regions compared to backward regions. All classes of households participate in the lease market both as lessors and lessees. However, while in the backward agricultural regions, the traditional pattern is more common wherein the small and marginal farmers dominate lease market as lessees and large and medium farmers as lessors; in agriculturally advanced regions, the lease market is in a state of transition where all classes of households participate. The trend towards reverse tenancy is more pronounced in all these regions. Among crops, the proportion of leased-in area is very high in the case of non-foodgrain crops compared to food crops.

Table 3.12: Classification of States According to Share of Marginal and Small Holdings in Area Leased-in Under Tenancy Contracts for Different Duration: Major States: 1991-92 to 2002-03

Share	*1991-92*	*2002-03*
	Less than 2 year	
> 75 per cent	Bihar, West Bengal and Jammu & Kashmir	Himachal Pradesh, Karnataka, Kerala,, Orissa, Tamil Nadu, Uttar Pradesh and West Bengal
50-75 per cent	Assam, Orissa, Himachal Pradesh, Kerala, Tamil Nadu and Uttar Pradesh	Andhra Pradesh, Bihar, Madhya Pradesh, Gujarat and Jammu & Kashmir
25-50 per cent	Andhra Pradesh and Madhya Pradesh	Karnataka and Maharashtra
< 25 per cent	Gujarat, Haryana, Karnataka, Maharashtra, Punjab and Rajasthan	Haryana, Punjab and Rajasthan
	2 to 5 Years	
> 75 per cent	Bihar, Himachal Pradesh, Kerala, Uttar Pradesh, and West Bengal	–
50-75 per cent	Assam, Orissa, Haryana, Jammu & Kashmir and Tamil Nadu	Himachal Pradesh, Jammu & Kashmir, Orissa, Tamil Nadu and West Bengal
25-50 per cent	Andhra Pradesh, Madhya Pradesh, Gujarat and Karnataka	Assam, Bihar, Kerala and Uttar Pradesh
< 25 per cent	Maharashtra, Punjab and Rajasthan	Andhra Pradesh, Madhya Pradesh, Gujarat, Haryana, Karnataka, Maharashtra, Punjab and Rajasthan
	More than 5 years	
> 75 per cent	Assam, Bihar, Himachal Pradesh, Jammu & Kashmir and West Bengal	Andhra Pradesh, Jammu & Kashmir, Karnataka, Orissa, Tamil Nadu and West Bengal
50-75 per cent	Kerala, Orissa and Tamil Nadu	Assam, Bihar, Himachal Pradesh and Kerala
25-50 per cent	Andhra Pradesh, Karnataka, Madhya Pradesh and Uttar Pradesh	Madhya Pradesh and Uttar Pradesh
< 25 per cent	Gujarat, Haryana, Maharashtra, Punjab and Rajasthan	Gujarat, Haryana, Maharashtra, Punjab and Rajasthan

Recent studies have also shown that the small and marginal farmers lease-out land primarily due to an increase in the cost of production, growing scarcity of water, falling returns and increasing uncertainty on account of erratic weather conditions. There is also an anecdotal evidence to suggest that many farmers, including small and marginal ones, are leaving their land fallow in view of restrictive tenancy laws.

Insofar as terms of tenancy are concerned, share tenancy with input-cost sharing continues to be an important mode of leasing-in land, particularly for small and marginal farmers. Though output sharing ratios vary from state to state, most of the studies do report 50:50 shares or even less. Most of the studies also show that the area leased-in for growing non-foodgrain crops is mostly under fixed money or fixed produce. Further, while in agriculturally-backward regions, share tenancy with input-cost sharing is more common, in agriculturally—developed regions, fixed rent tenancy is more pronounced. The micro studies also seem to suggest that small and marginal farmers prefer to lease-in land under share tenancy with input-cost sharing, perhaps because of lack of resources to pay cash rent in advance under fixed money and their inability to bear entire risk of crop failure which has increased in recent times. Recent studies from Bihar, where tenancy relations are presumed to be more exploitative, have also confirmed the above findings on magnitude and terms of tenancy.

The studies examining the effect of tenancy on the inputs used and output produced have thrown up mixed results. While some studies actually find lower use of inputs and low level of yields on the leased-in plots compared to owned plots, the findings of others are just contrary to these. An exhaustive survey of literature, however, suggests that there is no conclusive evidence to support the hypothesis that yields under share tenancy are lower than under owner farming or fixed rent leasehold tenancy (Vyas, 1970; Rao, 1971; Rudra and Chakravarty, 1973; Otsuka *et al.*, 1992; Bhaumik, 1993). Likewise, there is also no conclusive evidence to indicate that the yield levels for households involved in interlocking of factor markets are lower than their counterparts not involved in interlocking (Sharma *et al.*, 1995). Further, the findings broadly support the hypothesis that households of different size categories participate in the lease market to utilize their indivisible and non-tradable inputs and capital resources like family labour, bullock labour, machinery (tractors, bore-wells), etc. more optimally and, therefore, reduce per unit cost of production (Bardhan, 1976; Laxminaryan and Tyagi, 1977a; 1977b; Bliss and Stern, 1982). Nonetheless, a variety of other factors like absenteeism of land owners, inferior quality of land, land not suitably located, escalation in the cost of production and growing uncertainty in crop production are also reported to be important reasons obliging landowners to lease-out land.

Conclusions

The evidence from NSS data at the state level has shown that during 1990s, incidence of tenancy in terms of proportion of operated area leased-in declined in most of the states. The proportion of households not operating land, a mixture of those who own but do not operate and those who neither own nor operate land, however, increased hugely practically in all the states, that is perhaps a manifestation of the agrarian distress. Although, the macro data at the state level provides no conclusive evidence to suggest that marginal and small holdings face discrimination in the functioning of lease market in comparison to their medium and large counterparts, the phenomenon of "distress leasing out" of land by small and marginal farmers is reported by the micro studies. There is also evidence to suggest that restrictive nature of tenancy laws have compelled many farmers, including small and marginal ones, to leave their land fallow and hence, forgo income from the land.

In view of the actual functioning of lease market suggested by available micro and macro evidences, there is a case for liberalization of lease market to activate the land market and improve the access of land to a typical Indian farmer who wants to overcome scale infirmities, especially in the interest of production and trade efficiency.

The existing tenancy laws should be amended to allow leasing-in and leasing–out land with the following conditions: (i) separating ownership and use rights, (ii) making ownership rights non-alienable and secure and use rights to be freely purchased and sold in the market, (iii) fixing the tenure of lease, (iv) recording of lease and more importantly, (v) allowing landowners to resume land for self-cultivation after the expiry of lease.

The legalisation of tenancy shall allow all sections of rural population to participate in the lease market depending on their resource endowment, with adequate safeguards to protect the lessees of lower categories. For example, in some states like Punjab and Haryana, the small and marginal farmers may be tempted to lease out and medium and large farmers to lease-in land. The studies have shown that large and medium farmers who had leased-in land from small and marginal farmers had been able to invest in modern inputs, reap the scale economies and raise their level of productivity. In a similar vein, the marginal and small farmers who had leased-out their land also gained in terms of their occupational mobility and realizing higher annual income. But under no circumstances, the lessees, particularly the medium and large ones, should be allowed to buy land from lessors belonging to small and marginal farm-size categories. This will remove the fear of small and marginal farmers being wiped out and their land getting alienated and transferred to medium and large farmers.

In other states like Bihar and Orissa, where wages are low and adequate employment opportunities are not around, the small and marginal farmers shall be in a position to enlarge their holding size and thus afford a reasonable level of living with all attendant benefits of tenancy like borrowing from financial institutional agencies. The medium and large farmers in these states are likely to migrate to urban areas to take non-farm employment opportunities without any risk of losing their land. In fact, legalisation of tenancy shall provide incentives to such farmer households to move out of agriculture without the risk of losing their land. Such households over the years, when their livelihoods become secure in the non-farm sector, are expected to sell their land. There is enough evidence to show that today's lessors in large parts of country are different from their predecessors in the 1950s, 1960s and early-1970s when many amongst them did not take any interest in the agricultural operations and were leasing out land to keep control over tenants by exploiting them. They take active interest in agriculture in terms of input-cost sharing and assume greater risk and uncertainty.

In sum, in the current context of agrarian distress, well functioning lease markets can contribute towards broad based rural development in several ways like facilitating land from less productive to more productive uses, encouraging rural households to take up non-farm jobs without the risk of losing their land and increasing investment incentives such that those who make such investment can enjoy the benefits even if they are not in a position to use the land personally. Legalization of tenancy is a better option not only to the Market-Friendly Land Reforms promoted by the World Bank (Deininger, 2003) but also to Confiscatory Land Reforms advocated by the so-called neo-populist school (Griffin *et al.*, 2002).

References

Bardhan, Pranab K. (1976), Variations in extent and forms of agricultural tenancy: Analysis of Indian data across regions and overtime, *Economic and Political Weekly*, XI (37-38): 1541-1546.

Bhaumik, S. K. (1993), *Tenancy Relations and Agrarian Development: A Study of West Bengal*, Sage Publications, New Delhi, 204 p.

Bliss, C. T. and N. H. Stern (1982), *Palanpur: The Economy of an Indian Village*, Clarendon Press, London, 340 p.

Deininger, Klaus (2003), *Land Policies for Growth and Poverty Reduction*, Oxford University Press, New Delhi, 292 p.

Griffin, Keith, Azizur Rahaman Khan and Amy Ickowitz (2002), Poverty and distribution of land, *Journal of Agrarian Change*, 2(3): 279-83.

Laxminarayan, H. and S. S. Tyagi (1977a), Tenancy: Extent and interstate variations, *Economic and Political Weekly*, XII (22): 880-883.

Laxminarayan, H. and S. S. Tyagi (1977b), Interstate variations in types of tenancy, *Economic and Political Weekly* XII (39): A77-A82.

Otsuka, Keijiro, Hiroyuki Chuma and Yujiro Hayami (1992), Land and labour contracts in agrarian economies, *Journal of Economic Literature*, XXX (December): 1965-2018.

Rao, C.H.H. (1971), Uncertainty, entrepreneurship and sharecropping in India, *Journal of Political Economy*, 79(3): 578-95.

Rudra, A. and A. Chakravarty (1973), Economic effects of tenancy: Some negative results', *Economic and Political Weekly*, VIII (28): 1291-1294.

Sharma, H. R., Parkash Mehta and Khetrmohan Mohapatra (1995), Tenancy relations and productive/allocative efficiency: A study in Orissa agriculture', *Indian Journal of Agricultural Economics*, L (4): 597-608.

Vyas, V.S. (1970), Tenancy in a dynamic setting, *Economic and Political Weekly*, V (26): A73-A76.

4

Changing Pattern of Risk in Agriculture: An Analysis of Rice Risk

B C Barah

Introduction

Over the decades, agricultural production system in India has undergone several structural adjustments. Notwithstanding the significant contributions of innovative modern technologies to increase production and accelerate productivity growth, the share of agriculture in national GDP has decreased from 50 per cent during the 1960s to 18 per cent in 2007. The favourable policy interventions and supportive services played a crucial role in improving performances and increasing household farm income. But, the externalities of intensive agriculture have adversely affected the quantity and quality of natural resources such as soil, water and biodiversity, which is detrimental to production. These, in turn, have caused downside trends in the growth of production and productivity.

The degradation of natural resources, climatic aberrations and market infirmity have increased the risk and uncertainty and unsustainability. The volatility in production and pattern of factor usage has also increased vulnerability. The risk and instability in production emerged as critical factors affecting the livelihood of the common people in the post-modern agriculture. The agriculture sector is also vulnerable to multiple sources of risks. The impact of risk has been examined on rice, which is the most important food crop and a source of livelihood of more than two-thirds of the population. Rice occupies the first position in total planted area at the global level.

Out of the 230 million tonnes of total foodgrain production in India, 96 million tonnes was of rice in 2007-08. Rice alone provides half of the total nutrition requirement of the consumers. The changing pattern of various components of income risk is a matter of policy concern.

Rice is a water-loving crop, but two-thirds of rice area is under rainfed production environment, wherein the green revolution failed to penetrate to improve production. In the rainfed areas, the inter-regional and inter-farm disparities in farm income are high, poverty situation is deplorable and more importantly, dependency on agriculture has increased in the rural areas. Under the circumstances, risk in rice is highly conspicuous and has emerged as a major source of instability in agriculture. In-depth analysis of income variability and factor shares in agriculture are required for proper understanding of the dynamics of farm economy.

The importance of rice for livelihood security of the rural population can be judged from the fact that 84 per cent of total rice farmers in India are small farmers, who produce rice mainly for own consumption. It is ironical that although, Indian agriculture is dominated by small farmers, they are deprived of access ro technology, information, production inputs and markets. Due to lack of effective strategies for technology dissemination and access, the probability of risk in agriculture has increased manifold. Thus, not only the growth has declined at a faster pace, the farmers are also confronted with high risk, uncertain production and farm income, and distorted household food-security.

Nonetheless, rice production system has witnessed dynamism in recent years. Figures 4.1 and 4.2 depict a typical pattern of change in rice productivity in eastern India over the 1980s and 1990s. The eastern India is characterized by low productivity and high production risk. The number of poor districts, defined as having rice yield of less than one tonne per hectare, reduced from 9 districts in the 1980s to three in the 1990s. During the same period, the number of districts having production of 2-3 t/ha increased from 3 to 53. It is notable that despite these positive changes, the overall dominance of uncertainty in production, income disparity and fluctuation have adversely affected household economy of the risk-aversed small and marginal farmers.

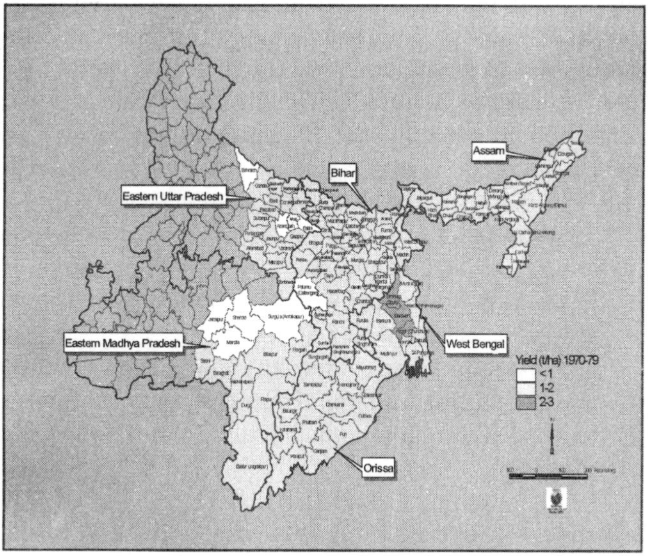

Figure 4.1: Spatial Pattern of Rice Productivity in Eastern India During 1980s

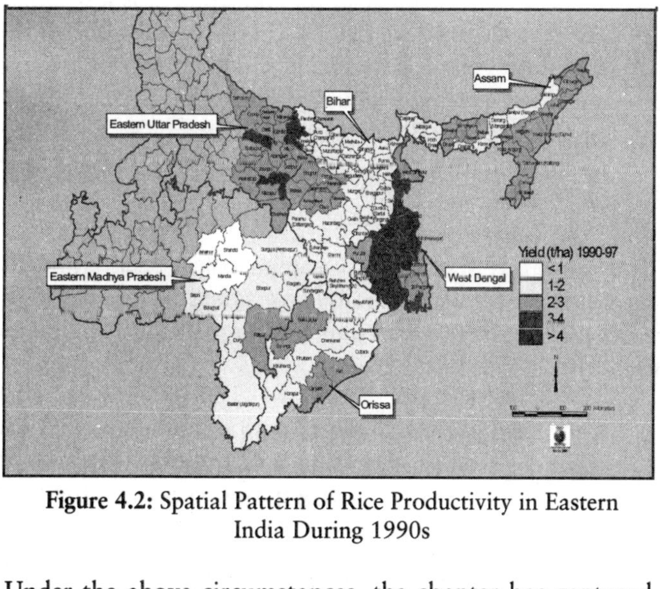

Figure 4.2: Spatial Pattern of Rice Productivity in Eastern India During 1990s

Under the above circumstances, the chapter has captured the nature and extent of risk in agriculture, particularly in rice production.

The bio-physical and socio-economic factors have been identified, which have differential impact of various components of total risk across the spatio-temporal dimensions. The components of total risk have been assessed to capture the dynamics over time and space by using the decomposition model. The chapter has focused on the following issues:

- To assess income risk and role and significance of its various components in agriculture,
- To decompose income variability into market component and biological component of risk,
- To analyze the factors affecting yield risk and price risk and assess their spatio temporal changes, and
- To draw inferences and derive policy relevant for risk mitigation.

In order to achieve the goals, a set of hypothesis was formulated; viz.

(a) The crop productivity in the rainfed area is low as well as associated with high risk,
(b) Risk in agricultural production and productivity has increased over time and spread across the regions, and
(c) Magnitude of yield risk is higher in comparison to price risk in the resource-poor areas, which is relatively high in resource-rich areas.

The role of risk in agriculture is important in decision-making at the farm level. Therefore, ideally, assessment of farm level risk is more relevant but due to data limitation, the analysis was restricted to risk at the district level.

- A time series district level data on rice production and other related variables was collected from the Directorate of Economics and Statistics for the past 3 decades for 13 states of India, and
- Information about modern rice varieties was collected from the Directorate of Rice Research, Hyderabad.

Analytical Tools

Decomposition of product variance following Taylor series expansion was used to measure the variability of crop income and of its components across agro-ecosystems (Goldberger, 1984, Piggot, 1978, Barah and Binswanger, 1980, 1982). The model decomposes the total variability into its components (price, yield and price-yield interactions). The method is generalisable, which helps in explaining regional behaviour in terms of various sources of total crop risk such as market risk and biological risk.

Decomposition Procedure

Denote ...... district "i" and year "t", where
i = 1,2, L (district subscript)
t = 1,2,....., T (year subscript)
Y = productivity per hectare (say Rice)
P = Farm Prices

The data series on Y and P are detrended appropriately to adjust for the technical change and inflation so that their variations reflect the pure risk only (Goldberger, 1970, Burt and Finley, 1964).

Let G be the unit gross revenue, then product of price and yield at time t is given by equation (1):

$$Gt = Pt.Yt \tag{1}$$

By suppressing the subscript "t" we may write the variance of G as

$$V(G) = p^2 V(Y) + y^2 V(P) + py\, Cov(PY) + R \tag{2}$$

Here, small letters denote the respective averages. Although variability of net revenue is most appropriate, the lack of required cost information restricts the analysis to gross revenue. We assumed that both revenue and cost behave identically.

For the sake of simplicity, a cv-type measure of V(G) was used as defined below to quantify income risk in each district. The individual components were standardized and expressed in percentage form, i.e.

$$CY\% = \frac{CV(Y)}{CV(Y)+CV(P)} \times 100$$

where

$$CP\% = \frac{CV(P)}{CV(Y)+CV(P)} \times 100$$

CV(Y) = Coefficient of variation (CV) of yield component,
CV(P) = Coefficient of variation of price component, and
CV(PY) = CV of interaction between P and Y.

Characteristics of Rice Economy In India

- Rice is a major staple food and most preferred cereal in the food basket. The total planted area under rice is around 44 million hectares in India, the world largest area under a single crop.
- Nearly two-thirds of total rice area of 26 million hectares is rainfed, where productivity is lower than the national average. The average yield of rice in the rainfed areas is < lt/ ha, ranging from half a tonne in parts of Madhya Pradesh to more than 4 t/ha in West Bengal. But, there exists a huge potential in the rainfed areas for increasing rice production. Even a modest target of about an additional tonne per hectare in the rainfed eastern India could increase national production by an additional rice of 26 million tonnes.
- High adaptation across diverse ecologies, inadequate infrastructural support and persistence of traditional technology are the major characteristics of the rainfed rice production system. These have resulted in below average performance, stagnating productivity and very low human development indicators.
- The preponderance of subsistence—oriented small and marginal farmers, having negligible marketed surplus, has perpetuated under-investment in agriculture. This has directly affected the performance in rice farming and in exacerbating poverty and hunger.
- Low productivity, high risk and uneven growth pattern have led to income losses. The near plateau of rice yield since the 1990s is a matter of concern, particularly when the population growth remains unabated. (Figure 4.3)

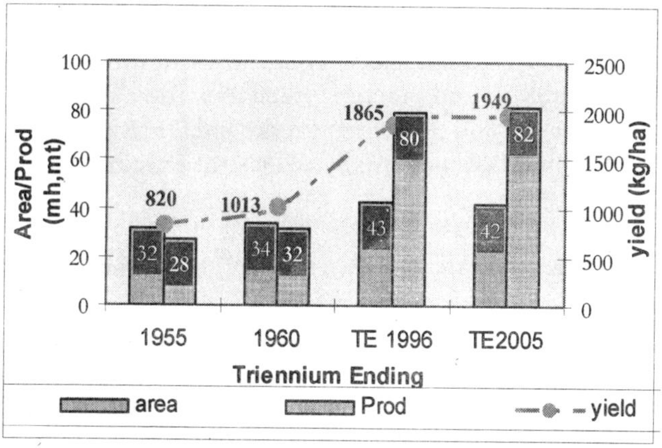

Figure 4.3: Rice in India

- Persistent low productivity has resulted in declining per capita availability of rice and has affected poverty as well as human health and nutrition. Figure 4.4 shows that the declining per capita availability of rice has reached a level below that of mid-1980s. In 2005, the average annual availability of rice was about 64 kg/person, which is lower by about 45 g/day/person, as compared to its highest availability of 222 g/day/person in 1991.

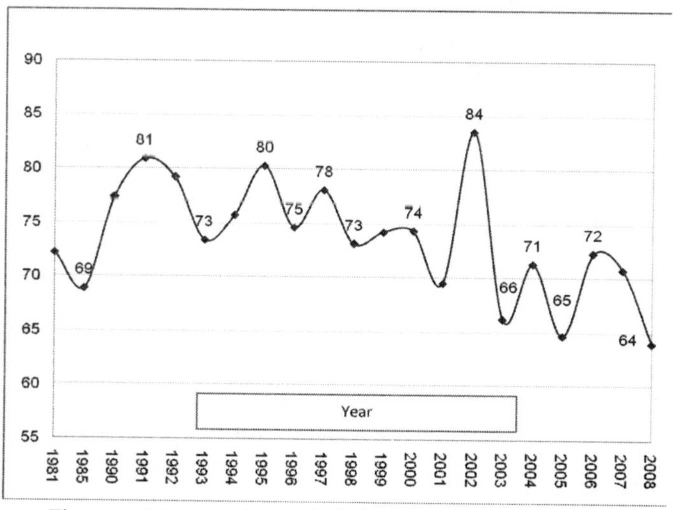

Figure 4.4: Per Capita availability of rice in India (Kg/yr)

- Growing vulnerability to climate change and biotic and abiotic stresses, are the big challenges in increasing productivity of rice. Therefore, adopting innovative technology and practices (such as aerobic rice, hybrid rice, SRI, etc.), and resilient to stress condition is utmost essential. The practices potentially reduce the existing yield gap to a great extent.

Factors Affecting Change in Rice Production System

1. *Nature of Irrigation:* Share of irrigated area under foodgrains increased from 24 per cent in 1970-71 to about 45 per cent in 2006-07, which has positively contributed to production increase. But, majority of the areas continue to suffer from unstable or scarce water resource in the rainfed areas.
2. *Adoption of Modern Varieties:* Adoption of newer high-yielding varieties (HYV) of rice had reached about three-fourth level in 1998 and remained sluggish thereafter. Lack of adequate irrigation facility has constrained adoption in the majority of the rice-growing areas.
3. *Price Volatility:* Since MSP (Minimum Support Price) is applicable to rice, its variability *per se* is not important. But, MSP is implemented only in a few states and in the remaining areas, price volatility prevails. At the same time, rising input cost has adversely affected the farm income.
4. *Shift in Rice Areas:* Area under rice has expanded to newer non-conventional areas and inter-seasonal shift is taking place from *kharif* to *rabi* season. Introduction of *Boro* rice as intra-crop diversification, has increased production of summer rice. But, irrigation in the *rabi* season is crucial for a *Boro* crop.
5. *Increasing Importance of Yield as Source of Production Growth:* In view of limited area expansion, breaking yield barrier assumes greater importance. The adoption of existing varieties needs to be enhanced to penetrate widely across the rice-growing areas, not to speak of the newer varieties, else the yield gap will remain and limit the growth.

6. *Regional Performance of Rice Yield:* Inter-regional differences in rice yield show wide variability. If the yield level across the region is equated to the national average, then there will be enough rice for every one.

Risk Perceptions

Baring irrigated area (about 30 per cent of total cropped area), rice production is basically dependent on the vagaries of nature, which made production more vulnerable to risk. It may be noted that over the years, the nature and importance of risk have changed in the rainfed areas. Interestingly, the resource-rich and assured irrigation areas are also facing newer types of risk in agriculture. Usually, the price/market risk is prominent in irrigated areas as the production system enjoys the benefits of protective measures, but the newer source of risk in production is arising due to soil and climatic aberrations.

Multiple Sources of Risk: The severity of climate—induced risk in agriculture (rainfall extremities, drought, flood and submergence), inadequate access to input output market (marketing and delivery system) and shocks, and technological and biological factors (technology adoption, incidence of biotic and abiotic stresses, declining response to inputs) are some of the critical factors. The emergence of newer sources of risk in the assured areas due to changes in temperature regime, declining rainfall, high intensity frost, flood, drought and storm has expanded the boundary of traditional areas. The price crash at the time of bumper harvest and steep rise in scarcity accentuate income risk.

Risk varies according to the ecosystems and climatic conditions; it is more relevant for rice because it is cultivated in highly diverse ecosystems and heterogeneous production conditions. Dominated rice ecologies comprise deep and shallow water, dryland and upland ecosystem. Table 4.1 gives the ecosystem-wise area and production shares, most of these are fragile ecosystems. Barring about 30 per cent area under rice is irrigated, the shallow, intermediate water regime and upland rainfed ecosystem covers more than 50 percent of total area, while rainfed dryland and deep water ecosystem accounted for the remaining 20 per cent area. Eastern India is the rice risk hotspot in the country on account of the existence of the largest ecologically diverse rainfed area, where productivity of rainfed rice is low and production risk is high.

Table 4.1: Rice Production in Highly Diverse Environment in India

Ecosystem	*Area Mha*	*All India, %*	*Eastern India, %*
Dryland	5.94	14	73
Deep water	2.43	6	88
Shallow & Intermediate	17.15	39	83
Rainfed Upland	4.60	11	99
Irrigated area	13.48	30	45
Total	43.66	100	

Inter–Regional Variations in Rice Yield

Studies have shown that inter-regional variations in productivity of rice are conspicuously high. Out of 534 rice growing districts in India, about 110 districts accounting to 27 per cent of the total area, have been classified as high performing districts (>2.5 t/ha), which contributes 43 per cent of production. But, the productivity in the bulk of area, namely 343 districts occupying 56 per cent area, is low and contributes barely 37 per cent of production (Table 4.2). The productivity is highly sensitive to climatic shocks in the rainfed areas.

Table 4.2: Classification of districts under different productivity groups during Ninth Plan (1997-98 to 2001-02)

Category of districts	*No. of districts*	*Area (Mha)*	*% share of all India rice area*	*Production (Mt)*	*% share of all India production*	*Productivity (kg/ha)*
High productivity (>2500 kg/ha)	111	12.50	28.0	38.30	43.9	3064
Medium productivity (2000-2500 kg/ha	80	7.74	17.4	17.30	19.8	2235
Sub-total	191	20.24	45.4	55.60	63.7	
Medium to low productivity (1500-2000 kg/ha)	100	7.18	16.1	12.76	14.6	1777
Low productivity (1000-1500 kg/ha)	165	12.71	28.5	15.36	17.6	1208
Very low productivity (<1000 kg/ha)	81	4.47	10.0	3.60	4.1	805
Sub total	346	24.36	54.6	31.72	36.3	
All India	537	44.60		87.32		1958

Source: Directorate of Rice Development, 2006, Govt. of India

Trends in Rice Production in India: Table 4.3 shows that over the past half a century, area under rice has increased from 32 Mha in 1950 to 44 Mha in 2006, while production has tripled, from 29 Mt to 91 Mt. But, the era of high growth has retarded abysmally after 1990s, as reflected in the low and declining growth rates. While the area stagnated with growth rate hovering within a thin margin around zero, the yield growth tumbled to one per cent after reaching a high of 3.19 per cent in the 1980s. Strikingly, the production growth has declined to less than half of the population growth rate, which is rather disparaging. Under the circumstances, a careful analysis of production performance is needed to evolve appropriate strategy for production stabilisation.

Table 4.3: Trends in Rice Production in India: 1950-2006

	Average			*Compound Growth Rates*		
Period	*Area, Mha*	*Production, Mt*	*Yield*	*Area*	*Production*	*Yield*
1950-1965	32.8	29.5	892.1	1.45	4.35	2.86
1960-1970	35.9	35.9	999.0	0.83	1.19	0.36
1970-1980	38.6	44.8	1156.2	0.88	1.90	1.02
1980-1990	40.5	58.2	1436.3	0.41	3.62	3.19
1990-2000	43.2	80.1	1852.0	0.67	2.02	1.34
2000-2006	43.2	91.1	2084.0	-0.62	0.42	1.05

Nature of Risk in Rice

For the analysis of panel data, the risk in rice production was classified into broad categories according to production environment and ecosystems such as irrigated system and rainfed system.

Pattern of Risk in Rice

Sources of risk	*Eco-system*	
	Irrigated	Rainfed
Price (market risk)	High	High
Yield (biological risk)	Low	Yield risk is higher than price risk

Generally, the magnitude of price risk is higher than of the yield risk, but the magnitude of yield risk dominates price risk in the rainfed areas. Conversely, yield risk is lower in the irrigated environment, where the protective measures and resource endowment are favourable to production.

Components of Risk: Using the decomposition model, various components of income variability were estimated. The term 'risk' connotes variability measured in terms of magnitude of temporal variability around long-term average. The data set used to measure the risk in production, price and yield, is smoothened by using a simple trend deflator to reduce the effect of outlier or sudden spikes. The disaggregate districts level data are analyzed to estimate the relative risk. Out of 375 districts considered in the paper, 205 were rainfed districts in the 1980s. During the 1990s, irrigation facility was provided to 12 additional districts, which affected the yield performance. It was observed that the price risk dominated the yield risk in magnitude, although the nature of risk has changed drastically during the past two decades. The spread and intensity of overall risk is apparent as the overall number of districts having yield risk higher than price risk, has increased to 127 in the 1990s from 70 in the 1980s. The price risk was higher in 305 districts in 1980s and 246 districts in the 1990s. The yield risk, as expected, has increased in both magnitude as well as spatially in the rainfed districts (Table 4.4). Within the rainfed areas, the number of districts where yield risk was more than price risk has doubled, from 38 in the 1980s to 83 in the 1990s. It indicates that increase in risk is more widespread in the rainfed areas, while the same has marginally changed in the irrigated areas. That is yield risk increased in 44 districts in 1990s as against 32 in 1980s. This implies that the price component is a dominant source of income risk in the irrigated areas, while it is the yield component which is a dominant in the rainfed areas.

Table 4.4: Changing Pattern of Risk in Rainfed and Irrigated Ecosystems (No. of Districts)

Ecosystem	*Decade*	*Higher price risk*	*Higher yield risk*	*Total districts*
Irrigated	1980-1990	138	32	170
	1990-2003	125	44	169
Rainfed	1980-1990	167	38	205
	1990-2003	121	83	204

It has been observed that the price risk is higher than yield risk across various ecosystems (59 to 81 per cent districts). But, the yield risk has spread to more districts in recent years (19 per cent in 1980s to 41 per cent districts in 1990-03). Therefore, an effective differential policy perspective is needed for mitigating risk, particularly in the yield risk dominated areas. Apparently, the well-endowed districts require a more efficient price stabilization policy, whereas, both yield as well as price stabilization policy is also important in the resource-poor regions. For the subsistence oriented farmers who produce food mainly for home consumption, the market pricing is less of an incentive in the yield risk dominated districts. The stabilizing production policies like crop insurance and/or other yield stabilisation policies are most essential for these groups of farmers.

In order to understand the overarching importance of risk and intensity of instability in rice production, district level disaggregate data indicates horizontal as well as temporal spread and magnitude of risk (both yield and price components). As mentioned earlier, both yield and price risk are spreading in space as well as time dimension, which imply that risk is emerging as on critical factor for household food-security.

Yield risk has been found to increase in 106 irrigated districts, but the same has been observed in a larger number of (141) districts in the rainfed areas. The corresponding figures for price risk are 135 and 144, respectively (Table 4.5). This indicates that yield risk has not only increased over time but also spread to more districts in the rainfed areas. On the whole, yield risk has increased in combined total of both the ecosystems in 247 districts and price component in 279 districts. The yield risk assumes greater importance in rainfed areas in the changing policy context.

Table 4.5: Number of Districts Showing Increasing or Decreasing Risk in Various Ecosystems

Risk	*Ecosystem*	*Increasing*	*Decreasing*	*Total*
Yield component	Irrigated	106	63	109
	Rainfed	141	63	204
Price component	Irrigated	135	34	169
	Rainfed	144	60	204

Figure 4.4 depicts the picture of spatio—temporal dynamics of price risk in Andhra Pradesh. The price risk decisively and distinctly shows an increasing pattern as reflected in the systematically higher values of price risk in the 1990s (Period 2) as against the same in the 1980s (Period 1). This indicates the growing importance of market

risk in the overall risk. Therefore, an effective market plays a crucial role in risk mitigation. In contrast, yield risk is more scattered and pattern is haphazard and devoid of any systematic pattern (Figure 4.5). Thus modelling the behaviour of yield is more complex, difficult to draw definite inferences and derive policy interventions on mitigating yield risk.

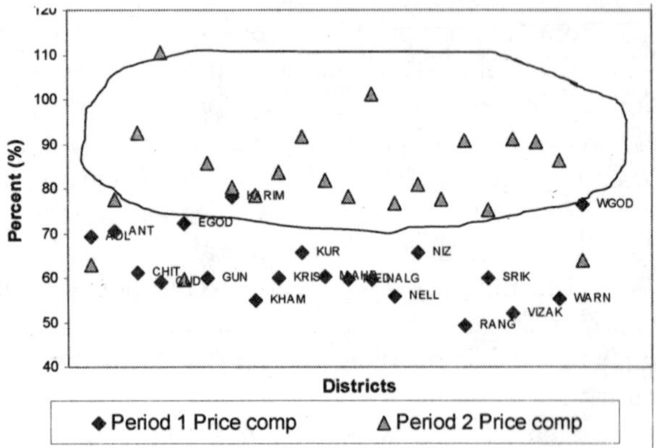

Figure 4.5: Changes in Price Component in Andhra Pradesh

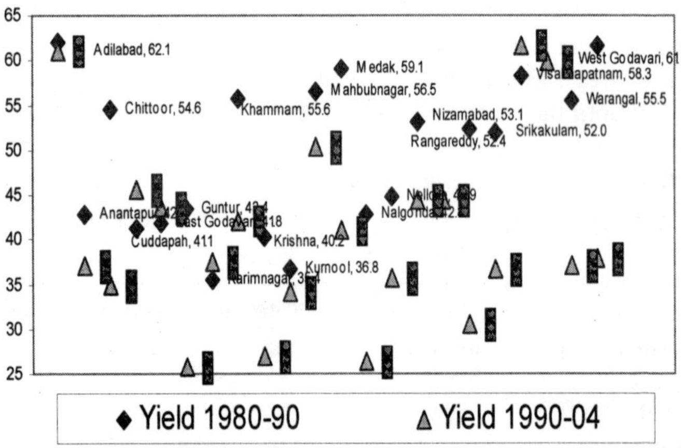

Figure 4.6: Changes in Yield Component of Rice Risk in in AP

Hence, there is a need for an in-depth examination of the cause and effect relationship of the risk dynamics. Table 4.6 integrates the input uses and the functional relationship of various inputs in relation to riskiness in agriculture.

Input-use Pattern: Wide variations in availability of inputs and their usages across the ecosystems influence the behavioural pattern of risk and its components. It is apparent from Table 4.6 that except in West Bengal, the average fertilizer use is lower in the rainfed areas by 20 to 95 per cent than in to the irrigated areas. Adoption of modern varieties is also lower in the rainfed areas, while in case of road density and market, the pattern of difference across the ecosystem is not discernible. It may be inferred that while the social infrastructures are necessary but the availability of improved variety increases the productivity. The inter-ecosystem variation in infrastructural development characterizes the visible difference in yield. Ironically, inspite of infrastructural development and higher input uses in the irrigated areas, the yield risk remains unabated due to lack of efficient management of resources and infrastructure.

Decomposition of Agricultural Risk

The total of 355 districts comprised of 205 rainfed districts and 150 irrigated districts. The total risk was decomposed into biological risk as represented by yield variation, and market risk represented by variability of price. The components of risk in the decomposition model indicate the relative importance of risk in agriculture in the changing context. The yield component is a major concern to the subsistence oriented small and marginal farmers who retain most of their production for own home consumption, more so in the rainfed areas. The production risk induces food insecurity, particularly at the household level, more importantly, when the spread and dimension of risk is changing over time and space (Barah, 2007). The emergence of newer source of risk in the irrigated areas despite effective technological and policy intervention are newer concerns. The estimated values show that as expected, irrigated districts experienced relatively less yield risk (53 per cent, as against 66 per cent in the rainfed areas) during the period 1980-90. In terms of spatial spread, it is clear that the yield risk has not only increased over time but has also spread to more districts in the rainfed areas. Yet, the risk due to market remains significant. The relative share of price risk was 69 per cent in irrigated areas during 1980-90, which has increased marginally to 71 per cent in 1990-03 (Table 4.7). The magnitude apart, the rate of increasing risk was more than the pace of decrease in risk. The relative yield risk reduced marginally from 66 per cent to 64 per cent, while price risk changed from 77 per cent to 64 per cent over the period respectively in the rainfed areas (Table 4.8).

Table 4.6: Characteristics of Rice Production in Rainfed and Irrigated Ecosystems in India

	Average Fertilizer use (NPK/ha)		*Average Road Density*	
	Irrigated	*Rainfed*	*Irrigated*	*Rainfed*
Andhra Pradesh	166.3	93.8	11.0	13.9
Bihar	94.3	36.4	7.1	12.9
Haryana	114.8		4.3	
Karnataka	130.2	75.0	24.0	15.0
Madhya Pradesh	53.8	29.6	3.3	4.0
Orissa	40.8	14.7	26.6	25.5
Punjab	161.4		7.4	
Tamil Nadu	148.1	118.6	26.8	28.8
Uttar Pradesh	101.7	32.0	2.8	2.0
West Bengal	86.4	109.7	7.6	8.8
Uttarakhand	108.1	5.5	10.9	15.6
	Irrigation, %		**Number of markets**	
Andhra Pradesh	59.0	27.4	36.3	36.8
Bihar	80.4	13.2	7.8	6.6
Haryana	75.9		26.1	
Karnataka	47.2	21.7	19.5	21.9
Madhya Pradesh	49.7	20.7	9.0	14.0
Orissa	54.0	25.2	10.5	11.1
Punjab	91.5		57.7	
Tamil Nadu	40.7	22.6	14.0	24.3
Uttar Pradesh	71.0	31.2	13.0	10.8
West Bengal	57.5	13.0	10.8	31.9
Uttarakhand	60.5	12.9	18.0	6.3
	HYV paddy, %		**Rice yield (kg/ha)**	
Andhra Pradesh	90.5	77.0	2661.3	2248.4
Bihar	49.8	44.5	1102.4	824.4
Haryana	34.8		2039.5	
Karnataka	41.3	50.0	2303.1	1955.6
Madhya Pradesh	41.7	48.6	1149.0	680.4
Orissa	77.2	56.4	1483.3	1264.7
Punjab	92.6		3285.4	
Tamil Nadu	71.0	69.8	2282.2	2772.4
Uttar Pradesh	70.6	48.3	1153.6	737.0
West Bengal	48.4	31.8	1511.6	1306.4
Uttarakhand	62.7	29.7	1922.5	1145.7

Table 4.7: Changing Pattern of Risk in Rice Production System Under Irrigated Areas in India

	Yield risk (%)			*Price risk (%)*		
	1980-90	1990-03	Risk change	1980-90	1990-03	Risk change
1. Andhra	47	37	Decrease	65	79	Increase
2. Assam	28	37	Increase	87	84	Decrease
3. Bihar	38	82	Increase	75	38	Decrease
4. Chattisgarh		68	Increase		12	
5. Haryana	80	52	Decrease	40	87	Increase
6. Jharkhand	45	70	Increase	100	43	Decrease
7. Karnataka	120	51	Decrease	135	86	Decrease
8. Madhya Pradesh	50	57	Increase	83	74	Decrease
9. Orissa	61	44	Decrease	68	57	Decrease
10. Punjab	26	30	Increase	84	78	Decrease
11. Tamilnadu	54	73	Increase	48	65	Increase
12. Uttar Pradesh	54	38	Decrease	64	75	Increase
13. West Bengal	57	39	Decrease	51	95	Increase
Total	53	49	Decrease	69	71	Increase

On the whole, the price risk is higher than yield risk in irrigated areas and it is reverse in the rainfed areas. Therefore, once again, it may be inferred that rainfed areas require more interventions for production stabilisation, while irrigated areas, require interventions on market improvement as the production is assured.

Table 4.8: Changing Pattern of Risk in Rice Production System Under Rainfed Areas in India

	Yield risk			*Price risk*		
	1980-90	1990-03	Change	1980-90	1990-03	Change
Andhra	51	42	Decrease	60	86	Increase
Assam		46			81	
Bihar	37	85	Increase	74	40	Decrease
Chattisgarh	39	80	Increase	84	47	Decrease
Haryana	90	34	Decrease	18	104	Increase
Jharkhand	42	79	Increase	77	40	Decrease
karnataka	118	62	Decrease	97	76	Decrease
Madhya Pradesh	63	72	Increase	92	69	Decrease
Orissa	63	58	Decrease	71	47	Decrease
Punjab	37	18	Decrease	84	101	Increase
Tamil Nadu	64	54	Decrease	46	71	Increase
Uttar Pradesh	69	50	Decrease	50	62	Increase
West Bengal	50	44	Decrease	69	90	Increase
Total	**66**	**64**		**77**	**64**	

Factors Affecting Agricultural Risks

Three broad categories of factors affect risk in agriculture, viz., Technological factors, Bio-physical factors and Socio-economic factors. Among the technological factors, modern high-yielding varieties are prominent.

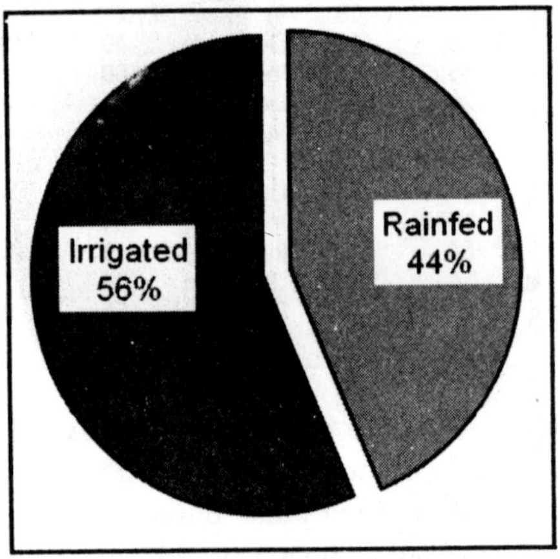

Figure 4.7: Availability of Modern Rice Variety by Major Eco-system (1965-2006)

Varietal development: The record of past four decades shows that as many as 720 modern rice varieties were developed in India. These varieties, however, are unevenly distributed favouring the well-endowed irrigated areas. More than half of the total number of modern varieties (MVs) are targeted to irrigated environment while unfavourable areas, which are highly diverse, get a lesser share, where as it needs the most. The unequal and slow adoption of variety affects the stability of the performance of rice. Obviously, an equiTable distribution of the HYVs across various ecosystems would reduce the inter-regional as well as inter-farm disparities in agriculture considerably. In fact, not only the availability, technology transfer is essential for a wider impact.

Biotic Factors: Incidence of pest and diseases hinders production and results in yield loss to the extent of 20 to 60 per cent, depending on the intensity of incidence. The occurrences of pest and diseases in rice have increased over the years, in both number as well as intensity.

In 1965, only three types of pests were found in the rice fields. Their number increased to 13 in 2000. Thus, the random but increased intensity of occurrence of pest and diseases has exposed agricultural production to more uncertainty (Table 4.9)

Table 4.9: Changing Pest and Disease Scenario in Rice Production Environment in India: 1965-2000

1965	*1975*	*1980*	*1985*	*1990*	*1995*	*2000*
Number (3)	*Number (5)*	*Number (8)*	*Number (9)*	*Number (11)*	*Number (13)*	*Number (13)*
Stem borer	BPH	BPH	Stem borer	Stem borer	Stem borer	Stem borer
Gall midge	Stem borer	Stem borer	Gall midge (biotype 4)	Gall midge	Gall midge (Biotype 5)	Gall midge (Biotype 6)
GLH	Gall midge	Gall midge (3 biotypes)	Leaf folder	Leaf folder	Leaf folder	Leaf folder
	GLH	GLH	BPH	GLH	BPH	BPH
	WBPH	WBPH	WBPH	BPH	GLH	GLH
		Cut worm	GLH	WBPH	WBPH	WBPH
		Gundhi bug	Cut worm	Cut worm	Cut worm	Cut worm
		Hispa	Gundhi bug	Gundhi bug	Gundhi bug	Gundhi bug
			Hispa	Hispa	Hispa	Hispa
				Thrips	Thrips	Thrips
				Mites	Mites	Mites
					Root weevil	Root weevil
					Black bug	Black bug

Source: DRR, Hyderabad

In addition, fragmentation of ownership holdings, growing family size, declining total factor productivity, imperfect market causing income fluctuation and more importantly, the overall effect of all these has resulted in reducing interest in agriculture. The findings of the recent NSSO farm survey indicate the exclusion theory. As a result, the agrarian distress has become a national phenomenon, which can't be ignored as a local event.

Concluding Remarks and Policy Perspectives

The analysis has clearly brought out as follows:

- Modern agriculture is highly input-intensive and sensitive to external shocks and vulnerability. In addition to bio-physical factors, the rising input cost and negligible increase in output price have resulted in variability in farm income. The decline in total factor productivity and erosion of farm business income have affected household food-security. Increased vulnerability to biotic and abiotic factors and changing risk dynamics are adversely affecting farming decision, which exacerbated agrarian distress.

- The analysis of spatio-temporal data shows that between 1980s and 1990s, the yield risk increased in intensity as well as spread in larger areas across the regions. The relative risk of both yield and price is high in the rainfed regions, although the yield risk dominates. Issue of instability in yield in the rainfed areas is a major challenge before biological researchers and policymakers. The growing yield risk perpetuates under-investment in agriculture and the lack of protective measures causes loss of farm income, particularly in the rainfed areas. Therefore, there is a need for strong public support system and interventions to mitigate yield risk. Introduction of crop insurance, effective input-output and knowledge delivery mechanism is needed.
- The behavioural pattern of price risk is also identical in both the ecosystems, though its magnitude varies. While higher yield risk prevails in the rainfed region, the price risk is a dominating source of agricultural risk in the irrigated areas. Therefore, effective price stabilization and other market support policy are relevant in the production assured areas.
- Analysis has shown that risk has increased over the year as the percentage of districts affected by severe yield risk has increased from 19 per cent in1980s to 41 per cent in 1990-03. As the biological factors and production fluctuations influence yield risk, prioritization of the strategies for yield risk mitigation, including R&D, and other interventions is essential, in the rainfed areas. Strikingly, the yield risk has emerged significant even in the irrigated areas *albeit* lesser in magnitude. This indicates that the vulnerability due to newer sources of risk is an emerging phenomenon for ensuring food security, not only in the resource-poor areas but also in the traditional well-endowed areas.
- The combined effect of lack of technological alternatives and inadequate infrastructure has caused stagnation in productivity and has affected the growth adversely. Therefore, technological interventions, including the development of stress-tolerant varieties and an effective

transfer of technology (TOT) mechanism are required to break the yield barrier. This necessitates more targeted research and development on crop improvement programmes.

- *Modern Variety:* Access to modern technologies and varieties is essential for the sustained production. The adoption of modern technologies at present is iniquitous favouring the irrigated areas. In fact, the rainfed areas, which are susceptible to biotic and abiotic stresses, suffer due to lack of protective means, including fewer varieties resistant to biotic and abiotic stress conditions. The stress-tolerant crop varieties also help in addressing the vexed problem of yield gaps.
- Lack of modern varieties as well as inadequate irrigation facilities exacerbate agricultural risk; therefore investment on irrigation and watershed development prorammes is needed.
- To correct market distortions and undependable delivery mechanism for inputs (seed, fertilizer and other agro-chemicals), concerted efforts and interventions are required to be undertaken by the government.
- The existing governance and functioning of state policies has become counter productive in many areas and has increased the dependence of farmers on the state. Therefore, state policies should be target-oriented and realistic to ensure inclusiveness.
- *Production Condition:* Less endowed regions occupy a major share of planted areas, but contribute little to the total production (e.g. 60 per cent rice area in the eastern India produces less than 50 per cent of the total production). Therefore, as there is vast untapped potential in rainfed areas, appropriate intervention on infrastructural support and technology is required to improve production environment and technological options for breaking yield barrier.
- *Price Response:* Very often, productivity appears to be insensitive to rice price, which signifies weak linkage of pricing policy and market system. In fact, the scope and implementation of MSP should be widened to cover larger areas and provide income support to the millions of farmers in the country.

- *Land Reforms*: Apart from fragmentation of landholdings, the existing unproductive tenancy system is regressive; thus there is lesser incentive to adopt modern technology. As a pro-poor measure, the land consolidation policy has worked well in the eastern Uttar Pradesh, which has facilitated improving well irrigation system. There are ample evidences that operation Barga played critical role in agricultural renaissance and has increasee rice production in West Bengal.
- The findings of the analysis of spatio-temporal dimension of risk argue for developing and choicing regionally differentiated policy for risk mitigation in response to relative importance of various components of risks. Unless risk (yield risk in rainfed area and price risk in well-endowed areas) is managed by prioritization of risk hotspot appropriately, the situation will accentuate the farmers' distress, which has become more apparent in recent years.

The estimation of various components of overall risk is relevant for priority setting, policy designing and targeting future investment in the rainfed agriculture. Apparently, the hotspot of yield risk requires more urgent policy attention and carefully—designed interventions. The resource risk in the protective areas also requires attention and income stabilization policies. The rural infrastructure is directly related to agricultural production, although its impact varies across regions, thus prioritization and sequencing public investment and infrastructure are needed as per the extent of conceived risk.

Note

1. National Centre for Agricultural Economics and Policy Research, New Delhi 110 012, barah48@yahoo.com.

References

Arrow, K.J. (1970) *Essays in the Theory of Risk Bearing*, North Holland Publishing Co, Amsterdam, Chap 4 and 5.

Goldberger, A. S. (1964) *Econometric Theory*, John Wiley and sons Inc New York.

Burt and Finley (1964) Statistical analysis of identities in random variables, *Political Economy*, 50:734-744.

Barah, B. C. and H. P. Binswanger (1980) Yield risk, risk aversion, genotype selection; *Research Bulletin No. 3*, ICRISAT

Barah B. C. and H. P. Binswanger (1982) Regional effect of national stabilization policies, *A case of India; Journal article* No. 219 ICRISAT

Reddy Ratna, V. and S. Reddy (2007) Agrarian crisis and labour market dilemma in India; National Seminar on Agrarian Distress: Causes and Remedies, August 1-2, NCAP, New Delhi.

Barah, B. C, H.P Binswanger, NGP Rao and B. S. Rana (1981) Risk aversion in plant breeding, Concept and application; *EUPHYTICA* 30:451-458.

Piggot, R. P. (1978) Decomposition of Variance of Gross Income, paper presented at the 22nd Annual Conference of Australian Agricultural Economists Society, Sydney.

5

Agrarian Crisis and Labour Market Dilemmas in India

V. Ratna Reddy and P. Prudhvikar Reddy

Introduction

The increasing frequency of farmers' suicides has been a cause of deep concern for various sections of the society. These suicides have roots in the agrarian and rural crises inflicting the Indian economy. The contribution of agriculture is declining at a faster pace than the population depending on it. While the share of agriculture in GDP is about one-fourth, 58 per cent of the population still depends on agriculture. Agriculture recorded the lowest growth rate of 1.86 per cent per annum during the past decade (1995-96 to 2003-04) as against 3.33 per cent during the earlier period. The value addition from agriculture has also decelerated, particularly in per capita terms. Growth in per worker income in agriculture has declined from 1.16 per cent during 1988-89 to 1993-93 to 0.28 per cent during 1998-99 to 2003-04, while in non-agriculture sector it has gone up from 3.31 per cent to 4.30 per cent during this decade (Chand, 2006). On the other hand, growth rate in wage earnings was higher of rural casual labour than urban casual labour. Between 1999-2000 and 2004, the increase in wage earnings of male and female workers was 24 per cent and 23 per cent in rural areas as against 19 per cent and 16 per cent, in the case of urban areas respectively. Based on the trends in employment, income, and wages, some scholars argue that declining employment growth has adversely affected the agricultural labour (Patnaik, 2006), while others maintain that agricultural labour is relatively better off as compared to the farmers (Reddy, 1994; Reddy et. al., 2005; Sen and Bhatia, 2004). Further, it is also contended that the focus of agricultural policies is oriented towards keeping the poor and marginal farmers in agriculture rather than shifting them towards high productive sectors (Srinivasan, 2006).

Against this background, this Chapter has examined the viability of agriculture and the relative shares of farmers and labourers over the past three decades to understand the genesis of the agrarian crisis. Attempt has been made to examine whether the crisis is endogenous or exogenous to the agriculture sector. The stagnant and declining low land productivities, due to technology fatigue or human induced factors, like land degradation, etc., are a reflection of the endogenous nature of the crisis. On the other hand, the exogenous factors stem from policy distortions at the overall economy as well as sectoral levels, culminating in say, biased price policies in favour of certain crops or adverse terms of trade for the agricultural sector, etc.

The Database

The study is based on the cost of cultivation data for important crops from some selected states, covering a period of about three decades, from 1971-72 to 2002-03. Five important crops, viz. paddy, wheat, sugarcane, groundnut and cotton, spread over 10 states were selected for the analysis (Table 5.1). Selection of states was based on the area covered under a particular crop and representation of the region. Each decade was subdivided into three sub-periods, viz. early, mid and late and the triennium averages corresponding to each period were worked out in order to maintain consistency of the estimates. For example, the early-1970s denoted triennial average of 1971-72 to 1973-74; mid-1970s represented 1974-75 to 1976-77 and late-1970s meant 1977-78 to 1979-80. Similarly, the data was computed for 1980s and 1990s as well. For the current decade, only one sub-period, viz. early-2000s (average of 2000-01 to 2002-03) was considered. The entire data set was drawn from the published reports of the Directorate of Economics and Statistics. However, for some years, there was non-availability of data on cost of cultivation. In such instances, the averages of the available years of the triennium were calculated.

Different cost methods like C_2 and C_3 were used to estimate the net returns[1]. Net incomes of the farmers were calculated in money

Table 5.1: Selected Crops and States

Crops	States
Paddy	Andhra Pradesh , Tamil Nadu, Orissa, West Bengal, Assam and Punjab
Wheat	Punjab, Uttar Pradesh, Madhya Pradesh
Cotton	Andhra Pradesh, Tamil Nadu, Maharashtra and Gujarat
Groundnut	Andhra Pradesh, Tamil Nadu, Maharashtra and Gujarat
Sugarcane	Andhra Pradesh, Tamil Nadu, Maharashtra and Uttar Pradesh

as well as real terms, using the consumer price index for agricultural labour with 1971-72 as base. Similarly, money and real wages along with real incomes of labour were also calculated to compare the changes in their standards of living over the period of three decades.

Viability of Agriculture

Productivity Changes

During the past three decades (1971-72 to 2002-03), the yield rates of most crops showed a positive change. The magnitude of yield increase was as high as 5.39 per cent and 3.13 per cent in the case of groundnut and cotton, respectively, in the state of Maharashtra (Table 5.2). The yield of wheat and paddy had risen by 1.14-2.7 per cent in the selected states, while for cotton, groundnut and sugarcane, the increase was marginal.

There are notable decadal variations in productivity of the crops. The growth in yield rates has either stagnated or was marginal during 1990s/early-2000s for all the selected crops across the states. Several other studies, based on a wider coverage of crops and states also substantiated that in general, performance of agriculture had come down during 1990s as compared to that in 1980s, although there were wide variations in the performance of crops and regions (Sen and Bhatia, 2006). At the national level, the growth in yield rates for all the crops recorded its peak during the 1980s (1981-82 to 1989-90), while the post-1990s scenario was somewhat mixed (Bhatia, 2006).

Growth performance of all the crops was higher at 1.96 per cent during post-2000 period (2000-01 to 2004-05) as compared to that in 1990s and 1970s. Across crops, paddy and groundnut followed the same trend, while wheat and sugarcane recorded negative growth

Table 5.2: Per cent change in costs and returns of important crops during triennium ending 1973-74 to 2003-04*

Cost/Returns Structure	*Wheat*			*Paddy*					
	Punjab	*UP*	*MP*	*AP*	*TN*	*Orissa*	*WB*	*Assam*	*Punjab*
Cost C2	8.59	8.29	9.78	9.92	9.68	9.96	9.85	9.92	7.93
Yield	2.22	1.74	2.17	2.50	2.24	2.70	2.47	1.44	1.14
Gross value of output/ha	9.13	8.44	9.12	9.66	9.27	9.31	8.97	8.98	8.95
Net returns (FL. land, and capital)	9.57	8.75	9.67	10.02	9.94	10.11	9.70	9.67	9.60
Gross returns/cost C2	1.94	−12.08	−19.67	−19.16	−2.49	−25.69	−39.95	−29.24	59.54
Employment/ha	−2.26	−1.59	−1.01	0.13	−0.21	0.78	0.91	0.42	−3.06
Money wage	9.86	10.21	10.39	11.25	10.82	10.37	9.99	10.36	10.40
Real wage	2.00	2.50	3.27	4.17	3.81	3.23	2.73	2.81	2.78
Money income / ha over cost C2	11.25	9.14	5.98	2.84	6.88	7.53	7.75	5.64	9.92
Real wage income / ha	3.56	1.62	−0.69	−4.16	0.78	0.28	0.31	−2.32	0.91
Disposable income (current)	9.88	8.82	8.98	9.71	9.25	9.33	8.71	8.71	10.65
Disposable income (constant)	2.07	1.07	1.90	2.63	2.41	2.60	1.96	1.23	3.14
Triennium Price Indices	8.09	8.15	7.38	7.08	7.76	7.20	7.39	7.91	8.09

(*Contd.*)

	Cotton				Groundnut				Sugarcane			
	AP	*TN*	*Maha*	*Guj.*	*AP*	*TN*	*Guj*	*Maha*	*AP*	*TN*	*Maha*	*Guj*
Cost C2	5.75	9.42	11.51	8.61	9.25	9.52	9.17	11.54	9.05	10.15	7.65	8.63
Yield	-0.07	1.37	3.13	0.97	0.50	0.56	2.51	5.39	0.08	0.99	0.20	0.87
Gross value of output/ha	5.81	7.70	9.71	8.56	8.75	8.63	9.51	9.99	8.16	9.51	6.06	7.02
Net returns (FL, land, and capital)	6.38	9.85	10.53	9.51	9.23	9.55	9.68	11.26	8.55	10.96	7.37	8.79
Gross returns/cost C2	12.91	-36.27	-44.96	-28.21	-16.71	-43.5	11.4	-19.39 -	13.93	-15.87	-46.07	-5.5
Employment / ha	-1.78	1.65	1.28	0.34	-0.53	0.13	0.84	4.15	-0.58	0.11	-1.31	0.15
Money wage	11.02	10.97	10.48	10.52	11.26	11.0	9.39	11.00	10.97	12.22	10.15	9.84
Real wage	8.58	4.00	4.42	3.44	4.22	3.54	1.97	3.28	10.82	5.33	2.99	1.96
Money income / ha over cost C2	6.10	0.00	9.98	14.85	Neg.	*	10.76	21.29	4.28	8.13	3.05	5.55
Real wage income / ha	-5.52	-7.62	2.87	7.07	Neg.	*	1.75	12.37	-3.34	2.67	-4.31	2.34
Disposable income (current)	6.40	7.82	7.82	8.99	8.42	8.28	9.97	8.63	7.78	9.86	5.27	7.12
Disposable income (constant)	-0.52	0.53	4.18	4.200	1.37	2.23	2.05	3.01	0.15	3.00	-2.00	0.85
Triennium Price Indices	2.45	4.08	12.68	5.67	7.13	7.76	7.67	7.37	7.13	7.76	7.37	8.15

Note: See end note

*Data available only for early-1970s and early-1980s.

rates during the post-2000 period. But, the actual yield rates between 1998-99 and 2005-06 were either stable or recorded marginal changes (Bhatia, 2006). Several factors, such as saturation of the green revolution technology coupled with the degradation of land and water resources due to monoculture, intensive agricultural practices, over-exploitation of water, etc. have led to decline and/or slowdown in agricultural productivity.

Net Returns

As a result of intensification of inputs during the 1990s, rising prices of fertilizers, water charges and other inputs, there was a substantial increase in the costs of production; it was at a faster rate than the value of output (Table 5.2). In fact, gross value of output had grown at a faster pace than the cost C_2 in only five out of twenty cases studied. These included wheat in Punjab and UP; paddy in Punjab; cotton in AP; and groundnut in Gujarat. Gross return to cost ratios had either declined or stagnated in all the cases, except paddy in Punjab, cotton in AP and groundnut in Gujarat. The net returns were highly fluctuating and quite low in most of the cases. In majority of the cases, net income per hectare started declining by late-1970s or early-1980s. Only wheat appeared to be an exception, as it recorded increased net returns in the 1990s. In fact, Punjab and UP had recorded highest net returns per hectare during late-1990s, while they were stable in MP. In the remaining crops, the net returns were too meagre to cover the supervision costs of the farmers (10 per cent of C_2). By and large, even though the new cultivars like hybrid rice or new methods of cultivation (SRI) had enhanced land productivity, they had not contributed to the net returns or profitability of the farmer (Chengappa et. al., 2003; Reddy et. al., 2005). The evidence from the cost of cultivation scheme data showed that net returns were not consistent with the differences between cost of production and minimum support prices, indicating that the MSP was not realistic in incorporating the inflationary trends (Bhatia 2006) and considering cost C_3, the net returns turned negative in majority of the cases.

Trends in Agriculture Labour Markets

Real Wages

Notwithstanding the declining profitability of crop farming, the real wages have recorded a positive trend over the period of three decades for all theselected crops in the sample states (Table 5.2). Highest growth in real wages was observed in the sugarcane in AP (10.8 per cent) and the lowest in case of sugarcane in UP (1.96 per cent). Cotton has recorded substantial increase in real wages in all the four states, with AP recording 8.58 per cent increase. Across the states, AP, TN and Maharastra have recorded higher gains in real wages as compared to other states, especially Punjab. These results are in contrast to the reported declining trend in real wages based on some of the aggregate data sets from other sources (Deshpande, 2006). However, it should be noted that the cost of cultivation data used in this study followed a rigorous data generation process (cost accounting method), and hence, tended be more precise as it reflected field realities to a large extent.

Employment

With the rise in real wages, the labour employment per hectare has been, by and large, either stagnant or declined sharply, as in Punjab (Table 5.2). The labour absorption or demand declined sharply in Punjab for both, wheat (by 2.26 per cent) and rice (3.06 per cent). Andhra Pradesh has recorded a decline in labour demand in cotton (1.78 per cent), groundnut and sugarcane. Also, labour employment has gone down for paddy in Tamil Nadu and sugarcane in Maharashtra.

Real Wage Income

The above two indicators show opposite trends and hence if interpreted separately, would lead to erroneous conclusions about the labour market situation. An increase in real wages along with a sharpe fall in overall employment will not help in ameliorating the conditions of agricultural labour just as a static or even falling real wages may lead to improvement in the conditions of agricultural labourers if the rise in employment is enough to offset the fall in real wages. Hence, real wage income of the labour (labour employment × real wage rate) is expected to give a clear picture regarding the changes in the conditions of labourers over time.

The data has shown that despite declining or stagnant employment, real wage incomes of the labour had increased over the period in all the cases. The earnings from hiring out labour had gone up in all the cases. This suggests that agricultural labours were better off as far as their purchasing power was concerned over the period in most of the cases. These observations are striking in the sense that the levels of living of the cultivating class are declining in terms of their disposable incomes while the incomes of agricultural labour are increasing.

Relative Factor Shares

In view of the evidence of declining net returns in crop cultivation and rising real wage bill of agricultural labour, the changes in relative factor shares have been brought out more sharply in this section. The relative shares of land, labour, capital and profits were computed with respect to both, value of output and value added to output. In the earlier studies on this aspect, it was often not clear whether factor shares were used with respect to value of outputs or value added to output (Prahladachar, 1983), although this distinction was important as factor shares do vary in magnitude, if not in composition, depending on the concept of output used.

It was observed that the changes in the relative shares were moving in favour of labour, capital and land (Table 5.3). Over the period, labour (family + hired) gained absolutely and relatively in all the crops and states, except in the case of wheat in Punjab. Its share in gross value of output had doubled over the period of three decades in a number of cases. The shares of land and capital had also increased, but to a lesser extent. In fact, in some cases the share of capital had even declined. On the other hand, the share of the farmer (profits) had recorded substantial decline in most of the cases. Profits were negative in three of the four states for groundnut. The declining profits of the farmer were somewhat compensated due to the extensive use of family labour. The share of family labour in the total labour was between 40 per cent and 50 per cent in a number of cases.

As far as value addition to output was concerned, the share of capital and labour was higher in a majority of the cases, while share of land was higher than labour share in a few cases (Table 5.4). The share of labour ranged from 52 per cent in the case of cotton in Tamil Nadu to 8 per cent in the case of wheat in Punjab. When the capital was divided further into material inputs (fertilizers, pesticides, FYM, etc) and animal and machine power, the share of labour was greater than material inputs in all the cases, except in wheat. Invariably, farmers were in the red (negative returns) in the cases

Table 5.3: Factor Shares in Gross Value of Output

(*in per cent*)

	Factor shares of important input costs in gross returns									
Inputs	*Early-1970s*	*Mid-1970s*	*Late-1970s*	*Early-1980s*	*Mid-1980s*	*Late-1980s*	*Early 1990s*	*Mid-1990s*	*Late-1990s*	*Early-2000s*
Paddy										
Andhra Pradesh										
Land	28.60	27.14	26.93	27.40	NA	29.84	NA	30.14	30.18	30.04
Labour	18.00	21.40	24.47	23.76	NA	21.27	NA	31.42	31.30	32.20
Capital	24.37	30.60	33.13	32.00	NA	26.72	NA	23.92	23.78	27.19
Residual (Profit)	29.03	20.85	15.47	16.83	NA	22.17	NA	14.52	14.74	10.58
Tamil Nadu										
Land	25.94	11.73	24.12	24.59	NA	NA	NA	NA	20.04	22.26
Labour	25.17	21.51	21.90	22.94	NA	NA	NA	NA	32.43	32.39
Capital	25.97	26.82	28.59	27.19	NA	NA	NA	NA	26.41	29.56
Residual (Profit)	22.91	39.94	25.39	25.28	NA	NA	NA	NA	21.12	15.78
Orrisa										
Land	24.93	23.64	24.96	25.82	24.84	NA	24.88	25.79	25.62	27.51
Labour	20.53	23.67	32.97	26.16	38.80	NA	29.74	31.76	36.29	40.02
Capital	22.38	26.42	26.17	23.72	24.38	NA	20.15	21.18	22.34	32.86
Residual (Profit)	32.16	26.26	15.90	24.29	11.97	NA	25.23	21.28	15.76	-0.39
West Bengal										
Land	24.52	25.36	24.23	25.16	24.81	NA	24.99	25.06	24.91	24.96
Labour	21.79	23.04	31.10	27.86	27.55	NA	28.51	32.57	36.39	47.61
Capital	18.69	27.79	29.30	26.09	20.08	NA	19.88	18.05	20.43	32.24
Residual (Profit)	35.00	23.81	15.37	20.89	27.56	NA	26.62	24.32	18.26	-4.80
Assam										
Land	25.69	22.54	23.13	23.56	23.52	NA	23.60	23.58	21.13	23.89
Labour	11.25	13.16	23.84	16.06	23.47	NA	19.44	18.12	17.36	24.11
Capital	19.23	18.18	19.84	19.51	21.91	NA	18.56	19.03	18.54	26.55
Residual	43.82	46.12	33.20	40.88	31.10	NA	38.40	39.27	42.97	25.45
Punjab										
Land	NA	25.52	18.94	25.28	25.00	26.81	28.94	31.76	32.38	33.54
Labour	NA	10.41	30.94	23.38	25.50	22.49	18.24	20.19	18.36	16.04
Capital	NA	32.03	25.83	26.48	24.24	22.61	18.53	21.19	21.85	21.16
Residual (Profit)	NA	32.04	24.29	24.87	25.26	28.09	34.29	26.86	27.41	29.26
Cotton										
Andhra Pradesh										
Land	NA	28.74	29.40	NA	NA	30.53	NA	28.87	28.63	28.97
Labour	NA	11.31	10.69	NA	NA	17.33	NA	22.15	26.21	24.02
Capital	NA	47.08	37.32	NA	NA	43.55	NA	28.83	34.11	33.12
Residual (Profit)	NA	12.87	22.59	NA	NA	8.59	NA	20.16	11.05	13.89

(*Contd.*)

(*Contd.*)

Inputs	*Factor shares of important input costs in gross returns*									
	Early-1970s	*Mid-1970s*	*Late-1970s*	*Early-1980s*	*Mid-1980s*	*Late-1980s*	*Early-1990s*	*Mid-1990s*	*Late-1990s*	*Early 2000s*
					Tamil Nadu					
Land	NA	18.43	20.81	NA	NA	NA	NA	22.94	27.66	23.97
Labour	NA	15.59	13.17	NA	NA	NA	NA	38.87	44.51	47.80
Capital	NA	23.00	26.26	NA	NA	NA	NA	19.83	24.91	19.15
Residual (Profit)	NA	42.98	39.77	NA	NA	NA	NA	18.36	2.92	9.08
					Maharashtra					
Land	NA	16.08	15.91	16.38	NA	NA	NA	16.67	NA	16.67
Labour	NA	19.69	24.98	26.77	NA	NA	NA	29.86	NA	40.30
Capital	NA	25.76	22.81	40.58	NA	NA	NA	29.08	NA	56.55
Residual (Profit)	NA	38.46	36.30	16.26	NA	NA	NA	24.39	NA	-13.52
					Gujarat					
Land	NA	16.09	16.65	16.43	NA	16.74	16.57	16.85	16.70	16.76
Labour	NA	17.02	15.08	18.08	NA	20.02	20.45	18.99	22.40	37.98
Capital	NA	36.00	43.15	40.43	NA	31.00	24.77	22.30	25.23	43.07
Residual (Profit)	NA	30.89	25.12	25.07	NA	32.24	38.21	41.86	35.68	2.19
					Sugarcane					
					Anhdra Pradesh					
Land	NA	NA	28.43	29.51	29.67	29.84	29.50	29.83	30.96	29.85
Labour	NA	NA	18.57	22.08	17.03	23.04	26.27	26.03	28.12	31.24
Capital	NA	NA	19.57	18.66	15.28	22.82	15.50	11.53	17.15	19.85
Residual (Profit)	NA	NA	33.43	29.75	38.01	24.29	28.73	32.61	23.77	19.06
					Tamil Nadu					
Land	NA	NA	NA	16.44	NA	16.82	NA	16.15	17.99	20.41
Labour	NA	NA	NA	16.72	NA	20.14	NA	22.64	26.58	29.93
Capital	NA	NA	NA	15.51	NA	13.87	NA	11.78	12.68	11.78
Residual (Profit)	NA	NA	NA	51.32	NA	49.18	NA	49.43	42.75	37.89
					Maharashtra					
Land	14.76	14.85	15.37	15.70	15.97	15.96	16.18	16.63	16.67	16.67
Labour	11.37	12.98	13.29	21.76	19.80	19.72	23.98	23.99	21.55	28.71
Capital	16.56	16.16	16.94	24.57	18.92	18.07	20.93	22.50	23.14	32.79
Residual (Profit)	57.31	56.00	54.40	37.97	45.31	46.25	38.91	36.87	38.64	21.82
					Uttar Pradesh					
Land	25.50	7.98	19.79	16.10	17.87	NA	20.41	25.71	24.21	23.05
Labour	13.45	4.97	14.47	11.72	13.88	NA	18.79	19.62	19.29	20.41
Capital	20.33	5.13	15.03	18.59	13.98	NA	11.63	13.38	11.65	15.80
Residual (Profit)	40.73	81.92	50.70	53.59	54.26	NA	49.17	41.29	44.85	40.74

(*Contd.*)

(*Contd.*)

	Factor shares of important input costs in gross returns									
	Early-1970s	*Mid-1970s*	*Late-1970s*	*Early-1980s*	*Mid-1980s*	*Late-1980s*	*Early-1990s*	*Mid-1990s*	*Late-1990s*	*Early-2000s*
					Groundnut					
					Andhra Pradesh					
Land	NA	26.06	25.89	29.77	NA	NA	30.04	29.87	29.46	30.85
Labour	NA	20.65	21.01	22.19	NA	NA	20.16	27.71	34.42	36.16
Capital	NA	49.12	40.91	40.44	NA	NA	37.74	37.97	39.79	54.03
Residual (Profit)	NA	4.17	12.19	7.60	NA	NA	12.06	4.46	-3.66	-21.04
					Tamil Nadu					
Land	25.38	26.37	NA	24.05	NA	NA	NA	26.02	26.00	27.78
Labour	18.32	26.21	NA	17.84	NA	NA	NA	30.57	34.56	47.41
Capital	31.05	43.36	NA	31.47	NA	NA	NA	33.68	28.82	50.22
Residual (Profit)	25.25	4.06	NA	26.64	NA	NA	NA	9.72	10.61	-25.41
					Gujarat					
Land	11.29	16.28	16.18	17.30	16.54	NA	16.59	16.72	16.72	16.50
Labour	15.37	12.26	18.90	18.46	12.80	NA	17.19	18.93	20.77	15.85
Capital	36.42	41.02	45.28	45.84	32.61	NA	42.96	33.41	31.57	27.10
Residual (Profit)	36.91	30.44	19.64	18.40	38.05	NA	23.26	30.94	30.94	40.55
					Maharashtra					
Land	NA	NA	NA	NA	16.25	17.07	16.61	16.67	16.67	16.67
Labour	NA	NA	NA	NA	17.53	17.99	15.03	27.44	33.66	36.77
Capital	NA	NA	NA	NA	48.88	38.00	30.05	34.06	41.07	48.74
Residual (Profit)	NA	NA	NA	NA	17.34	26.94	38.32	21.83	8.60	-2.18
					Wheat					
					Punjab					
Land	26.33	26.64	25.48	24.49	26.20	26.46	29.47	33.68	32.34	32.55
Labour	14.58	16.22	14.93	12.33	13.65	12.05	12.10	13.81	11.40	8.14
Capital	27.28	31.51	35.99	34.77	31.23	30.08	24.31	23.81	21.76	27.14
Residual (Profit)	31.81	25.64	23.61	28.41	28.91	31.40	34.12	28.71	34.51	32.17
					Uttar Pradesh					
Land	22.16	20.45	21.29	19.90	18.90	20.36	NA	22.02	21.88	23.44
Labour	13.96	14.24	15.74	14.00	14.47	19.00	NA	14.59	14.61	15.20
Capital	25.84	34.66	38.93	37.72	33.47	39.53	NA	25.75	23.44	27.76
Residual (Profit)	38.03	30.65	24.03	28.39	33.16	21.11	NA	37.65	40.07	33.61
					Madhya Pradesh					
Land	24.48	24.56	24.35	24.80	24.86	NA	24.91	26.12	25.00	25.00
Labour	14.00	16.00	15.13	10.37	13.42	NA	12.48	15.57	15.88	15.46
Capital	25.93	30.95	24.74	23.86	27.03	NA	25.73	28.01	26.54	28.03
Residual (Profit)	35.60	28.48	35.78	40.97	34.70	NA	36.88	30.30	32.59	31.51

Table 5.4: Factor share in value-added output: TE 2003-04

(*in per cent*)

Input	*Factor share in different states*					
			Paddy			
	A P	*T N*	*Orissa*	*WB*	*Assam*	*Punjab*
Land	30.14	22.02	27.72	25.01	23.73	34.25
Labour	33.25	32.87	44.05	50.44	45.35	12.79
Capital	49.71	53.26	54.51	55.46	39.82	34.47
Profit/Loss	–13.11	–8.15	–26.27	–30.91	–8.90	18.49
			Cotton			
	A P	*T N*	*Maharashtra*	*Gujarat*		
Land	29.06	24.77	16.74	16.84		
Labour	28.70	52.46	33.82	40.68		
Capital	49.31	40.87	81.57	63.44		
Profit/Loss	–7.07	–18.10	–32.12	–20.96		
			Sugarcane			
	A P	*T N*	*Maharashtra*	*U P*		
Land	18.67	12.46	27.59	39.00		
Labour	33.89	33.15	32.54	21.31		
Capital	46.14	37.91	57.81	25.04		
Profit/Loss	1.30	16.47	–17.94	14.66		
			Groundnut			
	A P	*T N*	*Maharashtra*	*Gujarat*		
Land	31.56	28.11	12.23	23.13		
Labour	38.44	51.49	29.96	21.77		
Capital	75.32	83.20	33.88	69.53		
Profit/Loss	–45.32	–62.80	23.93	–14.43		
			Wheat			
	Punjab	*U P*	*MP*			
Land	33.08	23.58	25.04			
Labour	7.60	15.34	15.58			
Capital	31.78	33.99	33.23			
Profit/Loss	27.54	27.09	26.15			

Note: Land = Rental value of owned land plus rent paid for leased-in land; Labour = Family plus, attached plus hired labour; Capital = Expenses incurred towards material inputs plus bullock labour hired and owned plus tractor hired and owned.

where the share of labour was higher. This clearly shows that labour was gaining in relative as well as absolute terms. The share of profits in the value addition was positive in only 7 out of the 19 cases. Of these, 3 were in the case of wheat (all), 3 in sugarcane (AP, Maharastra and UP) and 1 in paddy (Punjab).

On the whole, the status of the farmer appears to be relatively better in Punjab and UP.

Do Farmers Pay More than Labour Productivity?

The movement of factor shares in favour of labour and against cultivator prompted us to examine the wage rate in relation to labour productivity. The high magnitude of productivity-wage (money) ratios suggested that what a farmer got from a labourer was not less than what he had paid to the labourer (Table 5.5). However, these ratios were declining over time and had come down by half for all the crops, except wheat. In Punjab, the ratio had gone up substantially over the period, which could be due to the higher labour productivity achieved through mechanisation.

This trend appears to be in the direction of narrowing down disparities between landed and landless, but it is not a healthy sign as far as overall growth is concerned. Unless the decline in inequalities is accomplished through pareto-optimal welfare maximization where the welfare of one individual is increased without foregoing that of others, the equalities thus achieved may not sustain and, in fact, may result in aggravation of the problem in the long-run.

Labour Market Dilemma

The continuation of relative gains of agricultural labour and losses of cultivators may prompt the larger farmers to push down the labour share through bringing in more labour-saving technology. This in turn, would result in a shift towards capitalist agriculture where capitalist farms alone can survive and relegate small and marginal farmers to the ranks of labourers, accentuating the absolute disparities in income between labourers and farmers in the long- run. Recent trends towards increasing casualisation of agricultural workforce may be an early warning signal. However, except in Punjab, labour-saving mechanisation (capitalist agriculture) has not led to any substantial decline in the employment in most of the cases, as seen earlier.

The mechanisation of agricultural operations is often a source of controversy regarding its impact on various aspects of farm societies. While the land saving green revolution technology has more or less saturated in terms of its spread due to water constraints, the spread of labour-saving technologies seems to be limited. The spread of tractor use is evident in most of the states, the use of more capital-

Table 5.5: Average Labour Productivity Money Wage Ratios

States	*Early-1970s*	*Mid-1970s*	*Late-1970s*	*Early-1980s*	*Mid-1980s*	*Late-1980s*	*Early-1990s*	*Mid-1990s*	*Late-1990s*	*Early-2000s*
						Paddy				
Andhra Pradesh	5.55	4.67	4.09	4.21	NA	4.70	NA	3.18	3.20	3.11
Tamil Nadu	3.97	4.65	4.57	4.36	NA	NA	NA	NA	3.08	3.09
Orissa	4.53	4.53	3.66	4.31	3.48	NA	3.82	3.32	2.75	2.36
West Bengal	4.59	4.34	3.22	3.59	3.63	NA	3.51	3.07	2.75	2.10
Assam	4.30	3.90	3.35	3.62	2.74	NA	3.02	2.46	2.83	2.30
Punjab	NA	4.33	4.93	6.57	5.27	6.17	6.71	6.10	6.93	7.34
						Cotton				
Andhra Pradesh	NA	8.84	9.36	NA	NA	5.77	NA	4.52	3.82	4.16
Tamil Nadu	NA	6.41	7.59	NA	NA	NA	NA	2.57	2.25	2.09
Maharashtra	NA	5.88	5.96	4.33	NA	NA	NA	4.47	NA	3.12
Gujarat	NA	5.87	6.63	5.53	NA	4.99	4.89	5.27	4.46	2.63
						Sugarcane				
Andhra Pradesh	NA	NA	5.39	4.53	5.87	4.34	3.81	3.84	3.56	3.20
Tamil Nadu	NA	NA	NA	5.98	NA	4.97	NA	4.42	3.76	3.34
Maharashtra	8.79	7.70	7.52	4.60	5.05	5.07	4.17	4.17	4.64	3.48
Uttar Pradesh	7.44	20.13	6.91	8.53	7.20	NA	5.32	5.10	5.18	4.90

(*Contd.*)

(*Contd.*)

States	Early-1970s	Mid-1970s	Late-1970s	Early-1980s	Mid-1980s	Late-1980s	Early-1990s	Mid-1990s	Late-1990s	Early-2000s
						Groundnut				
Andhra Pradesh	NA	4.84	4.76	4.51	NA	NA	4.96	3.61	2.91	2.77
Tamil Nadu	5.46	3.82	NA	5.61	NA	NA	NA	3.27	2.89	2.11
Gujarat	6.51	8.16	5.29	5.42	7.81	NA	5.82	5.28	4.82	6.31
Maharashtra	NA	NA	NA	NA	5.70	5.56	6.66	3.64	2.97	2.72
						Wheat				
Punjab	6.86	6.17	6.70	8.11	7.32	8.30	8.26	7.24	8.77	12.28
Uttar Pradesh	7.16	7.02	6.35	7.14	6.91	5.26	NA	6.86	6.84	6.58
Madhya Pradesh	7.14	6.25	6.61	9.64	7.45	NA	8.01	6.42	6.30	6.47

End Notes:

Percentage change pertains to:

Wheat = Punjab – Early-1970s to Early-2000s; UP – Early-1970s to Late-1980s and Mid-1990s to Early-2000s; MP – Early-1970s to Mid-1980s and Early-1990s to Early-2000s.

Paddy = AP – Early-1970s to Early-2000s; TN – Early-1970s to Early-1980s and Late-1990s to Early-2000s; Orissa, West Bengal and Assam — Early-1970s to Mid-1980s and Early-1990s to Early-2000s; Punjab — Mid-1970s to Early-2000s.

Cotton = AP – Mid-1970s to Late-1970s and Late-1980s to Early-2000s; TN – Mid-1970s to late-1970s and Mid-1990s to E-2000s; Maharashtra – Mid-1970s to Early-1980s and Mid-1990s to Early-2000s; Gujarat – Mid-1970s to Early-1980s and Late-1980s to Early-2000s.

Groundnut = AP – Mid-1970s to Early-1980s and Early-1990s to Early-2000s; TN – Early-1970s, Mid-1970s, Early-1980s, Mid-1990s to Early-2000s; Gujarat – Early-1970s to Mid-1980s and Early-1990s to Early-2000s; Maharashtra – Mid-1980s to Early-2000s.

Sugarcane = AP – Late-1970s to early-2000s; TN – Early-1980s , Late-1980s, Mid-1990s to Early-2000s; Maharashtra – Early-1970s to Early-2000s; UP - Early-1970s to Mid-1980s and Early-1990s to Early-2000s.

intensive/labour-saving technologies such as harvesters/combines appears to be limited to Punjab and Haryana. Use of harvesters is confined mostly to paddy and wheat. The use of machinery in cotton/chilly picking or sugarcane cutting, etc., which are more labour-intensive operations, is yet to make a dent. While the absence of their adoption seems to be hurting the interests of farmers, their adoption may adversely affect the casual labour.

The flow of gains from agriculture is always uni-directional in a way that in the initial stages of green revolution, a major portion of its gains accrued to the cultivators (adopters) and in the latter years, they flew in the opposite direction cutting into the profits of cultivators. This is resulting in agrarian distress for the cultivators. The adverse effects on farm income (net) are seen in most of the states and crops where the spread of labour-saving technology is limited. This may lead to the argument that promotion of labour-saving technologies could reduce the agrarian distress. However, this would adversely affect the labour class, especially in the drought-prone regions, and hence, is unacceptable from overall welfare perspective. Thus, there is a need for a change in technology that would make cost-return ratios favourable without affecting labour absorption adversely.

Conclusions

Agriculture is becoming increasingly an unviable proposition. This is also reflected the National Sample Survey Organisation's 59th round survey, which reveals that one-third of the farmers have indicated that farming is not profitable and another 40 per cent of the farmers are ready to give up farming in favour of an alternative job. On the whole, 70 per cent of the farmers are frustrated with their profession (Deshpande and Prabhu, 2005).

This study has characterised agrarian distress by increasing cost of cultivation and decreasing farm profitability. Contrary to the general belief that increasing costs are associated with intensive input-use, i.e. material inputs, like fertilizers and pesticides, the study has linked the non-viability of farming to increasing labour incomes. There appears to be an inverse relationship between real wage income and net returns to farming. The results have revealed that agricultural labour is gaining while the farmers are loosing in relative terms.

Labour costs are rising due to higher wages as well as higher labour use (in a limited manner) per unit of land. One reason for this could be the increased interventions of labour-based programmes like watershed development, food for work, etc., which mandated minimum wage implementation. This trend is likely to continue, if not shifts upward, with the implementation of employment guarantee scheme[2]. Another reason for higher wages could be the increasing demand for unskilled (construction) labour in urban areas.

If this trend continues, the absolute disparities in income between labourers on the one hand, and owners of land and capital on the other may narrow down. But, in the light of farming becoming non-viable at a faster rate, consequent to increasing labour incomes, there is a possibility that small farmers may find hiring out rather than working on their farms more remunerative. Despite an uptrend in labour cost, the rise in adoption of labour-saving technologies to push down the labour share does not seem to be happening. However, the entry of large private enterprises in agriculture and resulting large scale investment in corporate farming, could lead to more and more use of capital-intensive agricultural technologies, thereby, pushing the actual farmers out of business and relegating small and marginal farmers to the ranks of labourers.

Labour-saving technology options are neither suitable nor sustainable in the absence of adequate alternative productive employment avenues to the rural labourers. This is more so in the context of dryland agriculture, where employment opportunities are limited and incidence of migration is quite high. Diversification to agricultural and allied activities that have the potential to generate for non-farm employment in the forward value chain could be a way out of the rural entrapment. While endowed regions could switch over to high-value horticultural/floricultural crops, fragile resource regions could move towards allied activities like low water intensive horticulture, dairy, poultry, etc. Also, switching over to cost-effective and ecologically-sustainable input-compositions that offset the increasing labour costs is another option.

For improving the viability of agriculture, access to complementary resources like water, credit, etc. needs to be improved. It is equally important that to check the agrarian distress, policy and institutional interventions address the existing distortions in factor and product markets. Appropriate policy support is needed to reduce the gap between producer and consumer prices. Providing infrastructure facilities that would ease the information and marketing bottlenecks can narrow this gap down. The recent initiatives of contract farming or the direct purchase of produce by the retail giants like, Reliance Fresh seem to be helping the farmers in getting better prices for their produce. However, appropriate policy and legal frameworks fostered with appropriate contract systems and enforcement mechanisms could only safeguard the interests of the rural communities at large. In the absence of policy and institutional initiatives in this direction, farmers would be left with no option but to hand over their lands to the corporate sector.

Acknowledgements: The authors thank Prof. C. H. Hanumantha Rao, Prof. D. Narasimha Reddy, Prof. R.S. Deshpande and Prof. S. Galab, for their constructive comments and discussions.

Notes

1. C_2= All paid out cost plus imputed costs of labour and land. C_3= Cost C_2+10 per cent of cost C_2 to account for managerial input of the farmer.
2. Preliminary field observations have suggested that the farmers complained of increased wages in the regions where the program was implemented in AP.

References

Bhatia, M.S. (2006), Sustainability and trends in profitability of Indian agriculture, *Agricultural Economics Research Review,* 19 (Conference No.): 89-100.

Chand, Ramesh (2006), India's agricultural challenges and their implication for growth and equity, paper presented in the *CESS Silver Jubilee Seminar on Perspectives on Equitable Development: International Experience and What Can India Learn?*, Centre for Economic and Social Studies, Hyderabad, 7-9 January.

Chengappa, P.G., Aldas Janaiah and M.V. Srinivas Gowda (2003), Profitability of hybrid rice cultivation: Evidence from Karnataka, *Economic and Political Weekly,* 21 June.

Deshpande, R.S. (2006), Agrarian distress and wages, Paper presented at the *48th Annual Conference of the Indian Society of Labour Economics,* Tiruchirapalle, 5-7, February.

Deshpande, R.S. and Nagesh Prabhu (2005), Farmers' distress: Proof beyond question, *Economic and Political Weekly,* XL (44 and 45).

Patnaik, P. (2006), The labour market under capitalism, *Indian Journal of Labour Economics,* 49(1).

Prahladachar, M. (1983), Income distribution effects of the Green Revolution in India: A review of empirical evidence, *World Development*, 11 (11): 927-944.

Reddy, V. Ratna (1994), New agricultural technology, increasing costs and relative shares of labour and other factors: A spacio-temporal analysis of Andhra Pradesh, *Indian Journal of Labour Economics*, 37(2).

Reddy, V. Ratna, P. Prudhvikar Reddy, M. Srinivasa Reddy and D. Sree Rama Raju (2005), Water-use efficiency: A study of system of rice intensification (SRI) adoption in Andhra Pradesh, *Indian Journal of Agricultural Economics*, 60(3).

Sen, A. and M.S. Bhatia (2004), *State of the Indian Farmer: A Millennium Study – Cost of Cultivation and Farm Income*, Vol. 14, Academic Foundation, New Delhi.

Srinivasan, T.N. (2006), *Trends in Employment, Unemployment and Wages in India: A Conceptual Framework and an Assessment of Findings*, Silver Jubilee Lectures, Centre for Economics and Social Studies, Hyderabad.

Section II

INTENSITY OF AGRARIAN DISTRESS

The section I brought out various problems faced by Indian agriculture. The policy apathy in dealing with the problems being faced by farmers has culminated in extreme distress situation and has manifested in the form of suicides by a sizable number of farmers. The increasing number of farmers' suicides highlights the fact that there is something fundamentally wrong with the agriculture, which has been ignored in India, particularly during the past one and half decades.

Distress is a manifestation of multiple causes, not necessarily related to extreme poverty, deprivation, and hunger. As discussed in section I, the agrarian distress has its roots in the problem of poor return from cultivation. The increasing dependence on purchased inputs has brought about greater demand for credit, adding another important dimension to the difficulties of the farmers as repayment of loans in the phase of low net returns in farming becomes unmanageable. The inability of non-farm sector to absorb the rural labour force in income generating opportunities accentuates the socio-economic malaise of rural India.

The extensive media coverage of farmers' suicides has brought the intensity of distress to the notice of policymakers. The phenomenon has been spatially widespread, and has not spared even the agriculturally most advanced state of Punjab. However, the worst hit states have been Maharashtra, Andhra Pradesh, Karnataka and Kerala, accounting for more than half of total farmers' suicides in country.

Section II presents detail studies on agrarian distress in two of these states, namely Karnataka and Andhra Pradesh. The study on Karnataka highlights the intensity of agrarian distress. Focuses on socio-psychological traits of distressed farmers committing suicides and considers indebtedness as the prime economic reason of distress and suicides in states. The study on Andhra Pradesh also takes a similar view about the causes of agrarian distress attributing it to the phenomenon of indebtedness due to demonstration effect. The policy response to address the agrarian distress has also been briefly discussed in the studies.

6

Farmers' Distress and Indebtedness: Miseries and Myths

R.S. Deshpande and Range Gowda

Introduction

Farming community is always exposed to high economic stress, compounded by social stress caused due to the collapse of village as an institution.The stagnating productivity of most of the crops together with marginal or no change in prices has telling effect on the net income of the farmers (Deshpande, 2005). The phases of stagnation that the Indian agriculture has gone through intermittently are perpetually weakening the welfare of vulnerables among the farmers, each time making them increasingly sensitive to the slightest economic shock, keeping them at the receiving end and facing net income crunch (Sawant, 1983). During the 1990s, the situation aggravated due to both policy failure and successive droughts at the end of decade. This period witnessed a phase of complacency and at the policy level, the problems of agricultural sector were taken for granted that led to slow agricultural growth (Bhalla, 2006).

The situation was quite disturbing in Karnataka, Andhra Pradesh, Maharashtra, Kerala and even the agriculturally most progressive state of Punjab (Reddy *et. al.*, 1998; Vasavi, 1999; Deshpande, 2002; Sainath, 2005; Mishra, 2005). In these states, agricultural production dipped significantly and the prices did not compensate for the loss in the gross income. On the other side, the cash component in the cost of cultivation increased, and as a consequence the net income flow to the farmer households stagnated. Farmers borrowed to meet the increased cost of cultivation, and other agricultural functions, but the shrinking net income constrained the payment of debt; thus, increasing the indebtedness of farmers' households. The spate of farmers' suicides that surfaced in these states was naturally associated with the performance of the sector, high incidence of indebtedness along with the other prominent factors like, advent of WTO, GM varieties, price collapse and spurious seeds. Though one cannot draw any 'one to one' correspondence between distress in the farm sector and the spate of suicides in some of the states, the farm and farm-related activities have a large stake in explaining the unfortunate occurrences.

This Chapter has attempted to understand the socio-psychological traits of distressed farmers committing suicides and has carried out detailed analysis of agrarian distress in Karnataka, focusing on indebtedness as the prime economic reason of distress and suicides in the state. Finally, the policy response towards the crisis has been outlined and suggestions for further strengthening the policy initiatives have been made.

Distress and Suicides: Socio-Psychological Traits and Empirical Evidences

Suicide is generally believed to be an impulsive act. But actually, it is a well-thought decision and has a cumulative causation. Psychology of the person plays an important role while taking any big decision in the life of a person. Psychologists provide scientific explanations to this phenomenon, whereas social scientists search for the socio-economic determinants. Confronting pure psychological theories, Durkheim (1952) had categorized suicides as *egoistic, altruistic, anomic and epidemiological*, based on the social response theory of distress. The egoistic person is more prone to suicide as the tolerance level is quite low and it is usually seen in the areas where semi-feudal production relations were prevalent. It is exactly the case of northern Karnataka, Telangana, Rayalseema (Andhra Pradesh) and Vidarbha region of Maharashtra. The *altruistic* suicides are common among the individuals who have inflated ambitions but unmatched capabilities. Here again, the emergences of production relations have a significant role. The *anomic* suicides are common among those who withdraw themselves from their group and here too, the ambitious farmers driven by the lure of quick and higher income become the victims. The circumstances provoke them to get on to a growth path, which is difficult for them to manage. According to some of the interpreters of Durkheim, distance of an individual from the society is considered as an important reason for suicide. The *Anomic* suicides are the result of a change in the social position of an individual due to any of the socio—economic reasons.

Another way of methodically analysing the causes of suicides requires understanding the incidence as culmination of four factors, viz. 'events', 'stressors' 'actors', and triggers. This categorization stems from the mental set-up of the victims (Deshpande, 2002). Among the 'events', crop loss, failure of borewell, price crash, daughter's marriage, family problems, property disputes, etc. get included. These become 'stressors' (stress creators) when two or more such 'events' cluster together. Specifically, prolonged illness of the individual or any of the family members, heavy borrowings, continued disputes in the family or land-related problems usually act as 'stressors'. These become lethal in combination with the 'events', but further ignition comes through the 'actors/catalysts' and 'trigger' incidence. 'Actors/catalysts' create a sense of 'insecurity' or 'insult' to the potential victim. These include the moneylender, banker, spouse, relatives, and close friends. Most often among relatives, the 'actors/catalysts' belong to the opposite sex of that of the potential victim. On the background of the 'events' and 'stressors', the 'actors/catalysts' fire the final shot by forcing an occasion to be the 'trigger' for the unfortunate incident. Given this complex nature of the phenomenon, it certainly becomes difficult to pinpoint one particular reason for the suicide (Deshpande, 2002).

According to the empirical information of the National Crime Records Bureau (NCRB), the suicide rate (number of deaths due to suicides per 100,000 persons) for male farmers in India has shown a rise from 12.3 in 1996 to 19.2 in 2004 and further to 18.2 in 2005. These rates for male non-farmers also increased from 11.9 in 1996 to 14.2 in 2000 and declined thereafter to 13.4 in 2005. Out of the total farmers' suicides in India, 89 per cent were males and 54 per cent belonged four states of Andhra Pradesh, Karnataka, Kerala and Maharashtra. The states with a high suicide rates among male farmers are: Kerala (195), Maharashtra (51), Chattishgarh (45), Karnataka (41), Andhra Pradesh (33), Tamil Nadu (32), and West Bengal (20) (Table 6.1).

Table 6.1: Average Suicide Rates for Male Farmers Across States: 2001-2005

State	*Farmers' group*	*Non-farmers*
Andhra Pradesh	33.2	20.6
Chhattisgarh	44.8	18.7
Karnataka	40.8	30.9
Kerala	194.7	50.6
Maharashtra	50.6	16.1
TamilNadu	32.0	24.7
West Bengal	20.0	20.4
India	17.5	14.2

Note: Calculations were based on suicides data by National Crime Records Bureau (NCRB) and Population data taken accordingly.

Source: Mishra (2005).

A survey in some of the distressed states plagued by farmers' suicides brought out multifarious reasons, incidents and events in the life of victims that eventually culminated in the form of such extreme action (Table 6.2). It was found that suicides are mainly attribuTable to social reasons such as family problems, old age and illness, and drinking and gambling habits. However, some rationalization is required to understand the reasons given by people who are close to the victim, as the victim's kith and kin may provide wrong information in the hope of getting some financial benefit from the government or they may sometimes conceal the facts fearing social insult.

The social and political relationship of the victims, their family commitments and support institutions assume great importance in their getting secluded and becoming introvert well before the occurrence of the incident. Durkheim's monograph on suicides indicates 'growing alienation of individuals from the family, society and religion' as the factor responsible for a suicide. Suicides could also be attributed to social disintegration. He has argued that more the family tries binding the individual to the domestic group, the greater is his social integration and less is the likelihood of his committing suicide. This proposition is somewhat corroborated by the information on the social and political relationship of the victims' families gathered during the survey.

In Punjab, it was observed that only 27 per cent of the victims were members of any political party and only 3 per cent of the members of victim families were the active members of the *Gram Panchayats*. The social relations maintained by the Andhra Pradesh

Table 6.2: Incidents and Events in the Life of the Victims in Different States

(in per cent)

Incidents and events from the life of victims	*Punjab*		*Andhra Pradesh*		*Karnataka*	
	Yes	*No*	*Yes*	*No*	*Yes*	*No*
Family and social problems	56.7	43.3	40.0	60.0	40.6	59.7
More alcoholism among other members of the family	3.3	96.7	6.0	94.0	6.6	93.4
Friction/quarrel about dowry with spouse/in-laws	–	100.0	10.0	90.0	4.3	95.7
Strong beliefs in rituals	3.3	96.7	2.0	98.0	8.7	91.3
Son/daughter left home	–	100.0	0.0	100.0	3.3	96.7
No child (specially a boy)	–	100.0	12.0	88.0	3.3	96.7
Daughter's/sisters marriage	3.3	96.67	20.0	80.0	21.5	78.5
Separated from friends	–	100.0	8.0	92.0	18.5	81.5
Huge expenditure/house construction	16.6	83.33	8.0	92.0	0.0	100
Problems from neighbour	–	100.0	2.0	98.0	11.5	88.5
Disputes with relatives	–	–	4.0	96.0	–	–
Chronic illness	–	–	10.0	90.0	–	–
Family member suffering from prolonged illness	–	–	6.0	94.0	–	–

Source: Coordinated Study on 'Market Imperfections and Farmers' Distress', ADRT Centre, Bangalore.

victims indicate that 34 per cent of the victims were interested in local politics. When asked about the guidance from others in the event of any problem, the family members gave favourable responses, i.e. the responses revealed that the victims had ample support from the Panchayat members, caste members, relatives and friends. But, a further probe and discussions about the behaviour of the individual led to the conclusion that most of the victims never discussed their problems with others.

In Karnataka, it was found that about 11 per cent of the members of the victim's family or the deceased persons had connections with the political institutions in the village. The participation in the rural political process makes a person more agile and enterprising. Withdrawal from such social processes definitely puts the person in a corner and generates the feeling of loneliness, neglect and insecurity. Therefore, the number of such persons in the victims was large who had no relationship with the political parties or had never participated in the process of rural politicization.

Social support system is one of the important factors that avert seclusion and introversion. The individual will not face an event of stress if he/she derives support from the family, friends and other social institutions. Such support will strengthen the belief in life and enhance entrepreneurial qualities. This also acts as a psychological insurance against any collapse of the person's faith in life. Family members, friends, relatives and village institutions, can provide social support to an individual. There was a general feeling that the social support has declined over time due to disintegration of the joint-family system and deteriorating standards of the village institutions, particularly, the cooperative culture. This can be one of the reasons for the spurt in suicides in recent times in some states, as hypothesized by Durkheim. A perusal of Table 6.4 reveals that the victims in Punjab received less support than required for the practical and emotional purposes. It was strange to observe that most of the families of victims were not in a position and therefore, not willing to support others in various practical and emotional activities.

Table 6.3: Social and Political Relationship of Victim's Families

(per cent to total response)

Sl. No. Type of relationship	*Punjab*		*Andhra Pradesh*		*Karnataka*	
	Yes	*No*	*Yes*	*No*	*Yes*	*No*
1. Was deceased belonging to any political party?	27	73	34	66	9	91
2. Membership of GP/TP/ZP?	3	97	8	92	11	89
3. Does your family spend any money on elections?	23	77	10	90	3	97
4. Is any member of family involved in betting, gambling, race, drinking, etc.?	43	57	34	66	24	76
5. Family commitments relating to:						
(a) Marriage/education of children	50	50	38	62	44	55
(b) Social functions	80	20			38	62
(c) Festivals	40	60	18	82	56	44
(d) Community programmes	7	93	22	78	15	85
6. Who guides you in your problems?						
(a) Neighbours	10	90	18	82	33	67
(b) Relatives	–	–	10	90	24	76
(c) Family members	90	10	38	62	45	55
(d) Panchayat members	0	100	60	40	11	89
(e) Caste members	33	67	42	58	–	–
(f) Others	17	83	14	86	17	83

Source: Coordinated Study on 'Market Imperfections and Farmers' Distress', ADRT Centre, Bangalore.

In Karnataka, the family and village institutions were found becoming more concentric. In other words, the support system that was being provided by the family and the village system in the earlier days was fast disappearing. This makes an individual, who is squarely caught in the web of the problems, bare open to the reality. If the events and triggers correctly match the time of the collapse in self-faith of that individual the decision towards drastic step is taken. It was located among the families of the victims that a good number of them (about 58 per cent) had mentioned about the requirement of support in day-to-day activities to the victim. Similarly, about 45 per cent had expressed the need of support even in the internal matters. Unfortunately, 34 per cent of them had indicated that no such expected support was received by the victim or his/her family (Table 6.4).

Table 6.4: Social Support as Reflected From Members of Victim's Family

Question	*Punjab*			*Karnataka*		
	No	*Little bit*	*Yes*	*No*	*Little bit*	*Yes*
1. In your opinion was he/she requiring any social support in						
• Internal maters	63.3	13.3	23.3	38.4	26.3	18.2
• Day-to-day activities	3.3	26.7	70.0	28.3	35.4	23.2
2. Do you feel that he/she was getting such support in						
• Day-to-day activities	36.7	15.5	46.7	34.3	31.3	20.2
• Internal activities	90.0	3.3	6.7	45.4	24.2	14.1
3. Do you feel others will get support from your family in						
• Day-to-day activities	60.0	13.3	26.7	40.4	32.3	12.1
• Internal matters	93.3	6.7	–	48.5	21.2	11.1
4. Do you feel you can support others in						
• Day-to-day activities	46.7	43.3	10.0	37.4	30.3	16.2
• Internal matters	76.7	13.3	10.0	43.4	21.2	11.1

Source: Coordinated Study on 'Market Imperfections and Farmers' Distress', ADRT Centre, Bangalore.

Besides the socio—psychological reasons discussed above, an important reasons cited in various studies associated with suicides was indebtedness. The borrowed funds were used largely for economically unproductive family commitments like marriage/ education of son or daughter, social functions, festivals, and community programmes. A detailed analysis covering this aspect, that is, indebtedness as the main reason causing distress has been taken up in the following section taking Karnataka as the case.

Indebtedness: The Raison d'être*—The Case of Karnataka

Changes in the pace and composition of agricultural credit have emerged as a significant issue in the context of the farmers' distress in various parts of the country. Karnataka invariably occupies a position close to the median in any comparison across states and farmers' indebtedness is not an exception to this pattern. During 1990s, farmers' debt burden increased substantially due to the changes in market situation, weather—inflicted uncertainties at the end of the decade and systemic flaws in the credit delivery system.

Acute indebtedness is arguably one of the prime causes of such a distress (Assadi, 1998; Revathi, 1998; Deshpande, 2002; Misra, 2005) that called for looking into the process of indebtedness and steps to correct the situation. In general, the credit flow to the agricultural sector is not meeting the demand for credit in the country and Karnataka is not an exception to this general pattern. In fact, the vulnerability is quite high in the case of Karnataka due to its high share of rainfed areas and weather–induced uncertainties.

The growth in total flow of credit (specifically agricultural credit) to various districts of Karnataka indicated that the total credit (including agricultural and other sectors) in the state increased at the rate of 15.6 per cent per annum during 1981-2000 (NABARD, 2006), though there were inter-districts variations. All the districts had experienced a reduced flow of credit to the agricultural sector during the 1990s, and that trend was slightly modified in the first five years of the present millennium. Not surprisingly, the backward districts of the state had received lower support as against the developed districts. It was noted that the ratio of agricultural credit to total bank credit (ACBC), in the state was 24.5 per cent during the period 1986-90 and declined to 16.9 per cent during 1996-2000 (NABARD, 2006). This suggested that the supply of agricultural credit as a proportion to the total bank credit was declining in Karnataka and sharply so in the backward districts of the state. Even though the proportion of agricultural credit was declining, banks had achieved their targets in all the districts, with a few exceptions, which pointed out to the faulty process of fixing targets. The problem of rural credit in Karnataka was largely due to the developmental disparity across the districts of the state. By categorizing the regions of Karnataka into three categories as developed, developing and under-developed districts, the large share of drought-prone area in the state almost covered the underdeveloped districts. It was not just a coincidence, but there was a logical reason of such perpetuation of underdevelopment in the drought-prone areas of the state. After every two years, the development clock in these districts gets reset by the climatic failure, and during every drought, the efforts of the state agencies are mostly focused on mitigating the immediate effects of drought. By the next monsoon season, all the alleviation measures are withdrawn or abandoned. As a result farmers are left to resurrect their farming activities with their own resources. Unfortunately, the borrowals during the previous year (depending on the severity and typology of drought), remain as overdue against the account of the farmer. Such overdue makes the farmer ineligible to get fresh advances from the bank. Consequently, a farmer (usually small and marginal farmers) is compelled to borrow from an alternative source of credit. Usually, the informal credit appears to have the advantage of easy availability and relatively low *apparent* transaction cost (at that time), even though the final cost of credit becomes huge. Table 6.5 depicts the sources of credit for the small and marginal farmers in Karnataka. About half of the credit need was met by the informal credit market as the time taken and the transaction cost of institutional credit was found comparatively high at the first instance (Datta and Sriram, 2003).

Table 6.5: Sources of Credit for Small and Marginal Farmers

Sl No.	*Sources of credit*	*Borrowers, %*
1.	Commercial banks and RRBs	29.54
2.	Co-operatives	13.81
3.	SHGs	8.82
4.	Total formal institutions	52.17
5.	Shops	8.08
6.	Chit funds	3.99
7.	Friends/Relatives/Moneylenders	35.75
8.	Total informal sources	47.82

Source: Based on primary data of a sample survey in Karnataka (Bhende, 2002)

In Karnataka, the problem of indebtedness is acute specifically in the rainfed regions of the state and also in the areas that have some potential for growth. Usually, farming community in poorly-endowed region does not undertake activities that require large investments and consequently, their credit needs as well as indebtedness are low. The farmers largely borrow from informal sources to avoid cumbersome documentation on one side and ease of obtaining the credit, on the other. At the same time, the developed regions and farmers seek larger quantum of credit and their resource as well as information base is good enough to earn better returns on the credit. Therefore, even though indebtedness as such is noticed among them but it is not chronic. The problem is, however, apparent in the regions that stand at the threshold of development. These regions as well as farming community have potential to grow. But, the farmers are unable to manage risks of the new vocations. The banking sector usually does not appreciate externality of the risk faced by this community of farmers and make provision to help mitigate its impact.

Services and Infrastructural Bottlenecks

High-tech agriculture has posed many new challenges to the credit sector. In this activity, the credit needs are comparatively high, and at the same time aggregate risk has increased. The credit provided by the cooperative sector is not sufficient, and there are no major schemes available in the cooperative sector to support such activities. NABARD refinances the cooperative banking institutions, and therefore imposes certain conditions for delivery and recovery of the credit. A branch of a cooperative bank is categorized as eligible/non-eligible based on its repayment performance. As the eligibility of the Primary Credit Cooperative Society depends on the repayment and the repayment performance of the Primary Credit Cooperative Societies in the underdeveloped regions is in general unsatisfactory. As a consequence over the years, these societies (in the underdeveloped regions), do not get adequate supply of credit and therefore, farmers from these regions have to depend upon the other informal sources of credit.

In the other formal credit source, viz. commercial banks and regional rural banks, the rates of interest are higher and documentation is cumbersome. Therefore, naturally credit from these institutions gets delayed due to the process of documentation apart from high transaction cost in the commercial banking sector. Besides, the availability of the branches of commercial banks in the remote areas of Karnataka does not meet the requirement in that region (specifically in Hyderabad—Karnataka region) (GoK, 2002).

Under the changed market situation, the farmer also requires guidance about the project and investment from the banking sector. On one hand, we find that the extension agencies are not fully equipped to provide guidance to the farmers to deal with the high-tech agriculture, and on the other hand, the commercial as well as cooperative banks do not have well-trained officers to guide the farmers. As a result, the farmer seeks requisite information from the dealers of the equipments and materials. The issue of these disadvantages confronting the farmers do not find place in the broader reform agenda and the farmers are constrained to depend on the exploitative institutions. This seems to have happened extensively in many parts of Karnataka.

The bottlenecks in the availability of the market infrastructure and functioning of the market have also perpetuated the distress of the farmers by shrinking the flow of income to the farm household. In Karnataka, the density of regulated markets and sub-markets is inadequate, and this has been pointed out as early as 1993 in Satish Chandran Committee Report (GoK, 1993). But, adequate steps were not taken to increase the density and/or efficiency of the regulated markets in the state. Further, the existence of crop-based specialized markets make it difficult for the farmer to take the crop to different places or sell it outside the market yard.

On one hand, low density of markets and the specialization limit the farmer's entry into the market to sell, while on the other, the middlemen offer all the services to the farmers readily inside as well as outside the markets. The process itself is such that the middlemen, as an institution are strengthened to exploit the situation. With inability of the credit institutions to meet the needs of farmers (needed quantity and time); the middlemen readily take advantage of the situation by providing credit to the farmers before the harvesting season and at times, even before the sowing season. This increases the dependence of farmer on the middlemen. Sensing better returns in this arrangement, the input dealers have also entered the credit market by providing inputs on credit to the farmers. This has not only facilitated increasing the total sales but has also helped in passing off the spurious inputs to the farmers and increasing their dependence on the specific input dealer. These two channels have strengthened the non-institutional credit market.

State Response to Farmers' Distress

Sensing the acute distress, faced by the farmers, the Govt. of Karnataka appointed a Committee under the Chairmanship of G K Veeresh (GoK, 2002b). The Committee recommended many interventions but faced rough weather due to the interpretation by the media. Following this, the Government of Karnataka had taken a series of steps to mitigate the distress of farmers in consultation with the members of the Committee. These are include:

- A health insurance scheme for the farmers, namely *Yeshaswini* was put in place, and the farmers have direct access to the best medical facilities available in the state-run hospitals. This has significantly reduced the expenditure of farmers on healthcare. The scheme has became quite popular among the farmers.
- As an immediate step, the interest rates on loans from Cooperative Banks were reduced to 6 per cent per annum and that had an important message to the farming community. It was further reduced to 4 per cent per annum in the 2007-08 Budget.
- The policy of providing compensation to the family of the victims (who had committed suicide) was thoroughly reviewed and compensation was allowed only in the cases where suicide was directly related to the farming activities. That had sent a clear signal across the state that compensation will not be provided automatically. A Committee under the Chairmanship of Sub-divisional Magistrate was appointed to investigate into every reported case of suicide and only on the recommendation of this Committee, the compensation was awarded.
- *Rayat Samparka Kendras* were provided Internet access and telephone facilities so that the initial signals of distress were transmitted directly to the state headquarters. Similarly, in the extension wing of the Department of Agriculture telephone links were established to reach the concerned official with ease.
- Crop insurance was taken up to *Hobli* level and compensation provided to the farmers based on the *Hobli* level data.

- A new scheme of seed subsidy was announced, and farmers were provided seeds with subsidy up to 50 per cent.
- The investment on watershed development programmes was increased by tenfold; it created increased employment opportunities in the rainfed areas.
- The 'Market Intervention Scheme' was reviewed and the corpus fund for the scheme was increased to Rs. 300 crore. Similarly, the market intervention scheme was made operational at the regulated market yard.
- Agricultural universities in the state were directed to focus on research and development responding to the demand rather than providing the 'supply-driven' research priorities.

Some other recent policy initiatives to mitigate the sufferings of farmers and address the issue of credit constraints included:

Kisan Credit Cards: Kisan Credit Cards have been introduced in the state as an instrument to increase timeliness availability of the credit. The policy, however, was half-hearted, pulled between two opposing objectives. On one side, the banking structure wanted the loans to be secured and released slowly, the farmers on the other side wanted the credit as and when needed, subject to the limit of their eligibility. With such interests at crossroads, the scheme seems to have got into a non-performing category.

A recent study by NABARD has brought out following points:

- Though there is a general awareness among the farmers about the Kisan Credit Card (KCC), there was low awareness about the special features of KCC, viz. flexibility in operation, reduced incidence of interest, provision for consumption loans, availability of personal accident insurance cover for KCC holders, etc.
- True benefits of KCC, however, have not percolated to the ground level since frequent operations in KCC accounts were restricted to 10-20 per cent of the total. Besides, the withdrawals were permitted only at the designated branches, which also acted as a limiting factor.
- Though there was less dependence of farmers on moneylenders after the introduction of KCC scheme,

borrowing from informal sources continued unabated mainly due to inadequate limit/non-availability of credit to farmers.

- Timely credit is not made available in Section 11 (1) to non-compliant DCC banks.

The scheme has not spread in the regions where it is needed most. The drought-prone backward districts of Karnataka are still to achieve sufficient spread of the KCC.

Relief Schemes: Most of the RRBs in the state have extended the facility of restructuring the loans under "Farmers in distress" and "Farmers in arrears" to provide relief under the Central Government's package on "Doubling of Agriculture Credit". Considering the consistent emphasis on financing of small and marginal farmers right from the inception of the crop loan policy, for the year 2006-07, the banks are required to ensure that at 30 per cent of the crop loan outstanding is availed by such class of farmers and 2 per cent of total crop loan outstanding is availed by the tenant farmers and oral lessees, who generally have limited access to credit. It was not implementable in Karnataka as there is a legal ban on tenancy.

Loan Waivers and Relief: In the Budget 2007-08, the Government waived the principal amount of outstanding loan to all farmers up to Rs. 25,000 and waived the interest to those who have borrowed more than Rs. 25,000, provided they pay the principal amount before 31st March 2008. While availing agricultural loans, farmers had to pay 1 per cent of the loan amount as registration charges; this additional financial burden on the farmer has been exempted for agricultural loans.

Table 6.6: Progress of 'Kisan Credit Card' scheme in Karnataka as on 31-03-2006

Agency	No. of cards issued during the year ('000)	No. of cards issued since inception ('000)	Out of which covered under PAIS ('000)	Aggregate limit sanctioned (in lakh Rs)
DCCBs@	113.4	1273.1.	906.8	368317
RRBs	113.9	696.7.	544.7	316046
Commercial banks*	221.2	1474.1	1135.1	679535
Total	448.5	3443.9	2586.5	136898

Note: @ Progress Report from 3 DCCBs not received.

· Information available upto Dec. 2005.

Source: NABARD (2006)

Exorbitant Interest Rate Act: The Government of Karnataka has enacted the "Karnataka Prohibition of Levying of Exorbitant Rates of Interest Act, 2004", to check levying of exorbitant rates of interest by private moneylenders. It has provision of taking stringent action against those moneylenders who violate the Money Lenders Act and levy exorbitant interest. Any such violation would attract imprisonment up to three years and a fine of Rs. 30000 or both.

Steps Towards Future Policy

The policy instruments addressing the issues of agrarian distress could be grouped into five important sub-systems, viz. those supporting (i) production sub-system, (ii) input sub-system, (iii) welfare sub-system, (iv) support subsystem, and (v) social sub-system.

The production and input sub-systems have to work through the extension network. We have failed miserably to maintain the tempo and penetration of the extension workers that was achieved during the phase of Green Revolution. Therefore, to help farmers on this front there is a need to build production capabilities and input delivery system that take care of reach as well as quality. It includes strengthening of *Raitha Samparka Kendras* (Centres providing information to the farmers), diversification of the farm activities, promotion of agro-processing, and up-gradation of technology in the input subsystem. The input-support system also requires information to the farmers on day-to-day basis through *Raitha Samparka Kendras*. It is for the first time that a welfare sub-system has been considered essential in this sphere. The Karnataka Agricultural Policy has recommended creation of a Farmers' Welfare Fund, a nodal department for the farmers and a wide social security system with facilities for the pension for the aged farmers. The support subsystem should include enhancement of facilities for healthcare (*Yeshaswini Scheme of Government of Karnataka)*, crop insurance and rationalization of the credit system. The social sub-system refers to the institutional structure at the village level. There is every possibility to rebuild the confidence of the farmers through formal and informal institutions at the village level. The role of farmers' organisations in this sector can be pivotal.

In the support sub-system, credit becomes an important component. It is necessary to revisit and rationalize the Figure of 18 per cent interest on credit to agricultural sector. Vyas Committee had expressed that the share of small and marginal farmers in agricultural credit should be commensurate with their holdings and credit needs.

This needs to be looked into seriously. Credit to small and marginal farmers should be progressively raised to 40 per cent of disbursements under the Special Agricultural Credit Plan (SACP) during the Eleventh Plan period. It is essential to integrate investment credit and production credit that will help in enhancing the capital formation. Further, considering the new development in technologies, scale of finance needs to be increased. It is also necessary that the collateral issue is revisited keeping in view the property structure and land records in the rural areas. The focus of the branch managers should shift from dependence on "Collateral Security" to "Creditworthiness" of the borrower and the "bankability" of the investment/activity.

Pledge financing, credit for marketing and introduction of Negotiable Warehousing Receipt System is required for enhancing the flow of credit. But, there is also a need of a long-term plan for development of waste and fallow land resources to bring them under productive use. As the drought is a regular phenomenon in one or other part of the state, therefore the schemes like (i) micro irrigation, (ii) sprinkler irrigation, (iii) watershed management, (iv) village ponds development, (v) farm ponds promotion, and (vi) dryland farming need priority in lending policy. The recommendation of Vyas Committee about setting up of an "Agri-Risk Fund" with equal contributions from the central and state governments and the participating banks needs to be taken seriously.

Note

*This portion is based on senior author's earlier work on farm sector indebtedness.

References

Assadi, M. (1998) Farmers' suicides: Signs of distress in rural economy, *Economic and Political Weekly*, 33 (14): 747-48.

Bhalla, G. S. (2006) *Condition of Indian Peasantry*, National Book Trust-India, New Delhi.

Bhende, M.J. (2002) Flow of Credit to Small and Marginal Farmers: A Case Study of Karnataka, Agriculturla Development and Rural Transformation Unit, ISEC, Bangkok, Ministry of Agriculture, Govt. of India, New Delhi.

Datta, Samar and M.S. Sriram (2003) *Flow of Credit to Small and Marginal Farmers in India*, Research Report, Centre for Management of Agriculture, Indian Institute of Management, Ahmedabad.

Deshpande, R.S. (2002) Suicide by farmers in Karnataka: Agrarian distress and possible alleviatory steps, *Economic and Political Weekly*, 37 (25): 2601-10.

Deshpande, R.S. (2005) Farmers' distress: Proof beyond doubt, *Economic and Political Weekly*, 29 October.

Durkheim, Emile (1952) *Suicide: A Study in Sociology*, translated by John A. Spaulding and George Simpson, Routledge and Kegan Paul, London.

GoK (Government of Karnataka) (1993) *Report of Expert Committee on Stagnation of Agricultural Productivity in Karnataka during 1980s* (Chairman: Satish Chandran).

GoK (Government of Karnataka) (2002a) *Farmers' Suicides in Karnataka–A Scientific Analysis*–Report of the Expert Committee for Study on Farmers' Suicides, Department of Agriculture, Govt of Karnataka, Bangalore.

GoK (Government of Karnataka) (2002b), *High Power Committee for Redressal of Regional Imbalances (HPCRRI)*, (Chairman: D M Nanjundappa), Govt of Karnataka, Bangalore.

GoK (Government of Karnataka) (2006) *Agricultural Policy of Karnataka–2006*, Department of Agriculture and Horticulture, Bangalore.

Mishra, Srijit (2005) *Suicides of Farmers in Maharashtra* (Interim Report), Indira Gandhi Institute of Development Research, Mumbai

NABARD (National Bank for Agricultural and Rural Development) (2001) *Report of the Expert Committee on Rural Credit*, NABARD, Mumbai.

NABARD (2006a) *Annual Credit Plan for Karnataka*, NABARD, Bangalore.

NABARD (2006b) *Submission by Dr D B Gore to the Committee on Marketing and Credit*, Govt of Karnataka, Bangalore.

NSSO (National Sample Survey Organisation) (2005) *Indebtedness of Farmer Household*, 59th round, Report No. 498, Ministry of Statistics and Programme Implementation, Govt of India, New Delhi.

Reddy, Sudarshan A., S. Vedentha, B. Venkateswar Rao, S.R. Reddy and V.Y Reddy (1998) *Gathering Agrarian Crisis—Farmers' Suicides in Warangal District (AP)-Citizens Report*, Voluntary Organisations, Warangal.

Revathi, E. (1998) Farmers' suicides: Missing issues, *Economic and Political Weekly*, 33(16): 1207.

Sawant, Shashikala (1983) Investigation of the hypothesis of deceleration in Indian agriculture. *Indian Journal of Agriculture Economics*, 38(4).

Sainath, P. (2005) Six out of ten?; Vidharbha: Whose suicide is it anyway? As you sow, so shall you weep, *The Hindu*, 31st May; 25th June; and 30th June.

Vasavi, A.R. (1999) Agrarian distress in Bidar, *Economic and Political Weekly*, 34 (32): 2263-8.

WHO (World Health Organisation) (2000) *Preventing Suicide: A Resource for Media Professionals*, Department of Mental Health.

7

Agricultural Distress in Andhra Pradesh: Causes, Consequences and Responses[1]

S. Galab and E Revathi

Andhra Pradesh had experienced a commendable breakthrough in the overall agricultural production and productivity for a short-spell from the late-1980s to early-1990s, preceding the beginning of reform era. There was a steep deceleration in the growth of both production and productivity in the state during the 1990s.The agricultural performance slipped much below the national average, leaving the state in a serious crisis.

This Chapter has taken an in-depth look into the root cause of the agricultural crisis in the state of Andhra Pradesh. More specifically, the study has addressed the factors contributing to deceleration in the agricultural growth in Andhra Pradesh, the consequences of the distress in agriculture and the responses of the state, market, civil society and the farm households to address the distress.

As Andhra Pradesh is not a homogeneous entity, but is consisted of three regions, viz. Coastal Andhra, Rayalaseema and Telangana, with widely different natural endowments, varying levels of infrastructural development, and undergoing differential long-term structural changes, the above issues have been addressed in the regional perspective.

Deceleration of Growth in Agriculture and Contributing Factors

This section is based on the secondary data, supplemented with the observations from the fieldwork.

Deceleration of Growth in Agriculture

During the 1990s, the agricultural growth rate in Andhra Pradesh was 2.2 per cent, which as noted earlier, was below the national average of 3.1 per cent. The share of income from agriculture declined steeply over time. The crop sector experienced a dismal performance with a very low growth rate of 0.59 per cent during 1993-94 to 2005-06. However, in some of the districts of predominantly dryland area of Telangana, high growth rates in agricultural output were experienced during 1991-92 to 2004-05. For instance, Warangal achieved a growth rate of more than 3.0 per cent per annum (Subramanyam and Aparna 2007). Interestingly, this district also recorded the highest incidence of farmers' suicides. This raises the issue about the cost at which the growth was achieved. The farmers in the dryland areas invest more in the groundwater resources to catch up with those in the irrigated areas. But, the failure of groundwater is a recurrent phenomenon in these areas. The cost of sunk capital is very high in the event of groundwater failure that at times culminates in loss of human lives (Revathi, 2007).

Contributory Factors

The factors responsible for the deceleration in agricultural growth included both, long-term structural and institutional factors and short-term, non-price as well as price factors.

Marginalization[2]

The landholding structure influences the productivity enhancing investments that have implications for agricultural growth. Over time, the institutional changes, market processes and demographic pressures have brought two remarkable changes in the landholding structure in A.P.; one, increased proportion of small and marginal holdings (accounting for 83 per cent of the agricultural operators) with almost half the area under cultivation and growing proportion of agricultural labourers. Second, there was a very high degree of tenancy, accounting for 50-60 per cent of the cultivated area. Much of this tenancy was oral, and unregistered, without any regulation and with exploitative rents, resulting in deleterious effects on agricultural productivity and growth. It was also pertinent to note that in the case of households belonging to scheduled castes and backward castes, accessing land through either 'sale-purchase market or lease-market' was on increase in recent times (Galab and Revathi, 2006).

The small and marginal farmers with relatively low access to irrigation facilities and no surpluses of their own have been getting into deep debt to augment investments in agriculture in the form of tapping groundwater resources. Thus, in the resource-poor dry regions or in the resource-rich delta region, a large proportion of "chronically disempowered" peasantry has emerged, who lack the wherewithal and incentive to make the necessary productivity enhancing investments (Reddy, 2007).

Stagnating Productivity and Production

Area under cultivation is an important determinant of agricultural growth. Over time, there has been a decline in the net area sown (NSA) and increase in the fallow land in the state. Though NSA in the coastal Andhra increased marginally, in Rayalaseema and Telangana areas, there was a decline in the NSA and increase in the fallow land. This indicates that the declining cultivated area had constrained the agricultural growth.

Besides declining NSA, the crop productivity also showed disquieting trends. Since the early-1990s, the deceleration in rice productivity growth has been much faster in AP than other major rice-producing states in the country. Growth of rice was tapered off by the low productivity level in the state. Groundnut is the major oilseed crop in the state in terms of production and area. Much of the area under groundnut crop is rainfed, and has been vulnerable to vagaries of monsoon. Hence, the level of production was highly volatile and showed a dismal performance in the 1990s. In the case of cotton crop, though, Andhra Pradesh had cotton cultivation under rainfed conditions, beginning late-1980, in certain dry regions there was a spread of cultivation of hybrid varieties of the crop under irrigated conditions. In the early-2000s, the area under cotton crop declined, but with a substantial increase in the rate of growth of productivity. Thus, deceleration in the growth of crops like rice, groundnut and cotton contributed for the overall poor performance of agriculture starting from the 1990s (Reddy, 2007).

Unstable Irrigation

Irrigation is another major driving force for agricultural growth. The actual area under canal and tank irrigation has been declining since 1990s with a phenomenal increase in the dependence on groundwater resources through wells and bore-wells. The more disturbing dimension in irrigation development of the state was that the groundwater sources in dry regions like Telangana and Rayalaseema accounted for 64 per cent and 70 per cent of the total net irrigated area, respectively. This means that farmers in dry regions bear substantial burden of irrigation provision from their own resources, often by heavy borrowings, and still end up with highly unstable and low-level of irrigation, while farmers in the relatively well developed Coastal Andhra enjoy the benefits of stable and high level of irrigation developed through public investment (Reddy, 2007).

The unstable growth of wells combined with monsoon failures and decline in surface irrigated area led to drying-up of wells due to inadequate recharge (Revathi, 2006; Subramanyam and Aparna, 2007)

High Cost Credit

The recent NSSO Survey (2003) has revealed that the farmers in Andhra Pradesh had low access to formal credit compared to all-India, and other southern states. Farmers from Andhra Pradesh had borrowed 68.6 per cent from the informal sources, while the corresponding proportion was only 42 per cent at the national level.

The bank credit disbursal also reflected the regional disparities in the state. In 2003-04, about 57 per cent of the bank credit to agriculture was to the relatively more developed coastal districts while the shares of backward Telangana and Rayalaseema areas were 29 per cent and 13 per cent, respectively. Cooperatives have been on the decline, RRBs have not been able to make any significant dent into rural financial needs and the commercial banks have been reluctant performers in agriculture with their increasing preference for lending to non-agricultural activities, as was revealed by the growth of urban branch banking. As a result, the farmers were forced to depend more on high cost credit for farm business investment and capital investment especially investment in bore-wells (Galab and Revathi, 2006; Subramanyam, and Aparna, 2007; Reddy, 2007).

The use of modern inputs like fertilizers, manure and improved seeds is, by and large, more extensive in Andhra Pradesh than rest of India. Usually, cash purchases are made for procurement of inputs, except in the cotton zone of northern Telangana. The formal institutions play important role in providing loans for cash purchases in south coastal Andhra and Rayalaseema. The situation was somewhat different in cotton and groundnut zones. During the focused group discussions in the villages, about 90 per cent farmers revealed that the new loans sanctioned by the formal institutions in the cotton zone were in the form of book adjustments, to repay the old loans. Similarly, farmers from about 75 per cent of the villages surveyed in the groundnut zone and 66 per cent of villages in the paddy zone reported book adjustments. Thus, the new loans sanctioned had not enabled the farmers to invest on the working capital. In addition to this, the loans from formal institutions had accumulated, as the old loans with interest were adjusted as new loans. As a result, the farmers were forced to approach informal sources like moneylenders, friends and relatives who charged very high interest rate. The limited credit flows from the formal institutions drived up the cost of cultivation, particularly for the marginal and small farmers as their access to the formal credit was less pronounced in all the zones.

The discussion with farmers on the conditions imposed by the informal institutions revealed that 'repayment of the loans is to be made immediately after the harvest of the crop'; thus, compelling the farmers to sell their output immediately and as a result they were not getting right price. Informal institutions like moneylenders, traders and friends and relatives had also mortgaged the assets of the borrowers. Considerable proportion of the villages also revealed that the 'private money lenders collect interest equivalent to one-year period even though the farmers repay the loans even before completion of one year'.

Declining Profitability

The data on Cost of Cultivation Scheme (CCS) at the state level clearly indicated a steep increase in the costs of cultivation at constant prices (base: triennium ending 1983-1984 prices), especially in the case of paddy and cotton during the 1990s, while there was hardly any corresponding increase in yield. The yield rates had either stagnated or increased marginally during 1990s and early-2000s in the state as a whole. Even the marginal improvements in yield rates during 1990s were achieved through higher input-use, resulting in faster growth of cost of cultivation than the value of output. It was also evident that the rate of change in money income over cost C2 from paddy was far less in Andhra Pradesh than other paddy-growing states, indicating the gravity of the situation in Andhra Pradesh (Reddy and Reddy, 2007). The cost of groundnut was around Rs 2620/ha during early-1980s, which increased to Rs 3920/ha in the year 2004-05—an increase of about 50 per cent. However, the rate of increase in the yields was far less (36 per cent) compared to costs at constant prices. By and large, similar increase in the cost per hectare of cotton at constant prices (an increase of 39 per cent) over early-

1980 was evident (Galab and Reddy, 2007). The increase in per-hectare costs indicated that the increased use of inputs did not commensurate with the increase in yields.

Consequences of Agrarian Distress

Farmers' suicides have to be considered as an extreme form of manifestations of agrarian distress. There are two interrelated processes that may lead to the painful act of suicide, viz. '*catching up process*' and '*indebtedness*'. The drive towards diversification, especially in the dryland areas in the central and southern regions of India, to crops like Bt cotton, was driven by the hope of farmers to increase yields and incomes. In the state of Andhra Pradesh, the farmers from the dryland areas of Telangana and Rayalaseema also diversified their cropping pattern more towards non-food crops. This process, called as '*catching up process*' leads to *indebtedness* in case the returns to investment are not in line with each other. The relatively low asset position of the farmers in dryland areas may be one of the reasons that force them towards higher dependency on non-institutional sources. However, creditworthiness is based mostly on social factors and recovery methods are more in the nature of social pressures than the legal processes (Reddy, 2007). This may in turn, create psychological stress for the farmers which ultimately results in the act of committing suicides. Hence, the analysis based on the primary studies was carried out and has been presented under three sub-sections, viz. '*catching up process*', '*indebtedness*' and '*suicides*'.

"Catching up" Phenomenon

A number of studies have addressed the issue of suicides in states like the Punjab, Karnataka, Maharashtra, Kerala, and Andhra Pradesh, locating the state-specific situation in the backdrop of changing macro policies. This indicates that the suicides were occurring in the high and medium growth states and they were conspicuously absent in the backward BIMARU states. Scholars have articulated high aspirations or the thrust for upward mobility, in the absence of public policy support, as the major cause for suicides in the backward areas of medium growth states (Rao, 2004). In order to examine this hypothesis, among others, a study was undertaken in the two districts of Telangana (Mahabubnagar and Warangal), and one district each in Rayalaseema (Anantapur) and o Coastal Andhra (Guntur).

Given the regional disparities in the development across the three selected regions of the state, the efforts made by the farmers in dryland areas of Telangana and Rayalaseema to catch up with the developed district (Guntur) highlighted interesting patterns. The cropping pattern across the four sample districts reflected the *catching-up behaviour*. The farmers in Warangal and Mahabubnagar districts had allocated the proportion of area to the non-food crops at par with those in the Guntur district. However, the farmers in Mahabubnagar had allocated considerable area under low-value non-food crops that required less intensive input-use. Interestingly, the farmers in the Warangal district had allocated the proportion of area for the cotton crop at par with those in the Guntur district. The farmers in Anantapur district dominantly allocated the area to one single commercial crop namely, groundnut. The efforts of the farmers in Telangana and Rayalaseema areas in following a cropping pattern similar to that in the agriculturally prosperous areas reflected their aspirations to catch up with the income levels of the farmers in the advanced regions.

The other indicator that reflected the catching–up behaviour was the irrigation base of the farmers, since irrigation improves the yield of the crops. The percentage of irrigated area to net sown area was found to be the same for the farmers in Warangal and Guntur districts, but the farmers in the Mahabubnagar were not able to catch up in this regard. This indicated that the farmers in the Warangal district were far ahead of others in the sample districts. But the farmers in the three sample rainfed districts had used groundwater resources to catch-up with those in the coastal districts. The entire growth in groundwater sources was due to private investment on these irrigation sources. Districts (Warangal, Karimnagar and Khammam) with high irrigation growth rates also had high agricultural growth rates, one reinforcing the other with the correlation coefficient between the two as high as 0.78 (Vakulabharanam, 2004). Minor irrigation census statistics showed that 89 per cent of the irrigation infrastructure in the state was owned by individual farmers, of which the share of Telangana farmers was 62.40 per cent (Third Minor irrigation Census, 2000-01). The ownership of irrigation wells and landholdings did not have strong correlations which meant that all categories of farmers were investing on tube-wells and it was more so by marginal and small farmers. Government ownership and investment on minor irrigation infrastructure, though meagre was, still concentrated in the coastal regions, despite Telangana and Rayalaseema deserving more attention.

Table 7.1: Incidence and Correlates of Suicides Across Different Districts

Sl No.	Variables	Warangal	Mahbubnagar	Anantapur	Guntur
1.	Percentage of villages affected by suicides	31.9	21.2	39.8	11.8
2.	Percentage of villages reporting one suicide from villages affected by suicides	67.7	78.6	79.6	84.7
3.	Percentage of villages reporting two suicides	18.2	18.0	13.9	12.9
4.	Percentage of villages reporting three suicides	6.7	2.4	4.3	2.4
5.	Percentage of villages reporting four suicides	3.4	0.3	1.2	–
6.	Percentage of villages reporting 5 and >5 suicides	4.0	0.7	0.9	–
7.	No. of cultivators (2001)(lakhs)	4.4	5.5	4.8	3.8
8.	No. of agri-lab (2001) (lakhs)	3.8	7.7	4.6	9.6
9.	No. of suicides (1998-2004)	689	387	501	147
10.	SMR (2004)	28.6	26.4	17.0	28.6
11.	Share of area under non-food crops to GSA (%)	41.8	62.6	83.8	23.7
12.	Share of area under cotton/groundnut/to area under non-food crops (%)	64.2 (cotton)	17.8 (cotton) 21.7 (groundnut) 25.4 (castor) 41.7 (other)	98.5 (groundnut)	69.6 (cotton)
13.	Area under food grains/pulses/chilli to area under food crops (%)	50.2 (food grains) 9 (chilli)	67.2 (food grain) 26.4 (pulses)	57.5 (food grain) 42.1 (pulses)	7.9 (chilli) 50.5 (blackgram)
14.	Ratio of agri. labourers to cultivators	1.2	0.67	0.74	0.32
15.	Percentage of marginal & small farm holdings	84.1	74.0	65.4	86.83
16.	Percentage of area under groundwater source to GIA	82.8	91.2	79.6	12.8
17.	Irrigated area to GCA (%)	54.1	16.8	12.4	53.1
18.	Area operated by marginal & small holders (%)	45.7	47.1	32.2	56.6
19.	Average size of holdings (ha)	1.85	1.44	2.00	1.05
20.	Average disbursement of crop loans per cultivator (Rs)	6474.46	4280.33	8727.84	18966.94
21.	Short-term loans (in lakhs)"	3023.45	3888.64	9386	16790
22.	Medium-term and long-term loans"	220.27	135.37	68.59 6686.55*	5586

Notes: *Disbursed under the head 'Others'; "Distributed by the DCCB, figures for year 2002-03.
Source: Handbook ofStatistics, DES 2004-05, for Warangal and Mahbubnagar and 2001-02 for the districts Anantapur and Guntur.

The incidence of suicides has been found to be the lowest in the coastal Andhra region among the three regions of Andhra Pradesh. Further, it was the lowest in the relatively backward areas of highly developed districts like Guntur compared to the dominantly dryland agriculture districts of Telangana (Warangal and Mahabubnagar) and Rayalaseema (Anantapur). Thus, the pattern of incidence of suicides indicates that the incidence of suicides is highly correlated with the pace with which the farmers in the backward areas try to 'catch up' with the farmers in the developed areas. The high incidence of suicides in Warangal, followed by Mahabubnagar from Telangana region has provided substantial evidence to this fact (for details *see*, Revathi, 2007).

Indebtedness among Farmers

The steep rise in cost of living in the rural areas is indicated by Consumer Price Index for Agricultural Labour (CPIAL), while the farmers' income languishes. It is a familiar scissors crisis of rising costs and declining returns in agriculture, often resulting in pauperisation of the peasantry. The NSS data has revealed that 82 per cent of households were indebted in the rural areas of Andhra Pradesh during the year 2003 (NSSO Report No.498, 59th round).

In a recently concluded study to assess the extent of indebtedness among farm households in the three zones (dominantly paddy, groundnut and cotton growing zones) of the state and across farm-sizes, the information on indebtedness was collected as on 31 May 2004 (Galab and Reddy, 2007). On an average, a farm household irrespective of the farm-size, was indebted to the extent of Rs 213887/-in the state of Andhra Pradesh. Across the zones, farmers in south coastal Andhra were indebted maximum to the tune of Rs 299418/-, followed by Rayalaseema (Rs. 198919/-) and northern Telangana (Rs 159788/-).

In all the zones, the indebtedness was positively related to the farm-size. For instance, a marginal farmer in the state was to pay the principle amount of Rs 130038/-during the reference period as against Rs 339715/-by a large farmer. The major sources of credit were moneylenders constituting around 29 per cent of the amount loaned (Rs 61500/-), followed by friends and relatives and traders with average loan amount of Rs 53905/-and Rs 31631/-, respectively. In all, about 70 per cent of the farm households were dependent on informal sources for their credit needs. It is well known that the informal sources, especially moneylenders charge higher interest rate than the formal institutions. The interest charged by informal sources was found as high as 33 per cent per annum, irrespective of the zone and class of farmers, whereas formal institutions charged around 10-15 per cent interest.Thus, indebtedness was widespread among the farming community irrespective of the differences in the resource endowment and ownership of land.

The purpose-wise analysis of credit during the five-year period ending 2003-04 revealed that on an average, 14-25 per cent amount borrowed was towards the fixed capital, mostly for irrigation (i.e. digging bore-wells). Nearly one-fourth of the amount borrowed in the groundnut zone, a dry zone (Rayalaseema), was found for fixed capital. If the expenditure incurred towards repairs was added, the share increased substantially. Another important expenditure was towards social events, such as marriages and death related expenses, especially in the paddy and groundnut zones. In the cotton and paddy zones, education and health related expenditure was higher compared to that in the groundnut zone.

Stress Factors Leading to Farmers' Suicides

An interesting debate on the role of stress or triggering factors in determining the incidence of suicides was centered on a basic question that given all the disadvantages common to all the farmers, why some among them commit suicides. A recently concluded study by Revathi (2007) has provided answer to this question. In her study, risks have been classified into co-variant and idiosyncratic categories. The co-variant risks include, natural disaster risks (drought, monsoon failure, cyclone hailstorm, pest attack) and output-related risks (crop failure, non-remunerative price, etc.), while the idiosyncratic risks are related to inputs (credit, irrigation, seed, pesticide), human capital (basic needs, healthcare, education), violence (family disputes, deaths, alcohol consumption), and human psychology (pressures for repayment of debt, stigma of loanes).

It has been found that for the suicide cases it were the idiosyncratic risks that were more fatal, although the empirical analysis has indicated that the incidence of suicides was influenced by both, the covariant as well as idiosyncratic risks. This gives an indication that the individual—related risks, which are person-specific, mattered the most, implying that such a state of affairs might not be easily amenable for policy interventions. Analysis based on further disaggregation of type of risks has shown that input-related risks and domestic violence-led risks had more influence on the incidence of suicides. In general, the input-related risks were high and uniform across districts, specially the risk of credit becoming debt was very high and almost universal in the study area.

It is also relevant to point out that input-related risks were connected to social identities of the farmers. The dependence of scheduled tribes on the markets for the inputs was less than that of other social groups. Hence, for STs, these risks may not be a serious problem now, as they are not completely integrated with the input markets. But, these risks were more pronounced among scheduled castes as they were completely integrated with the input markets. The SCs had weak bargaining power with the market forces and hence they encountered more risks. The input markets seemed to be perfect but were being operated within the framework of power relationships.

In some cases, however, the input-related risks were spread uniformly across farmers of all the social groups, as happened for BT cotton seed. Due to non-regulation, farmers in the cotton-growing district of Warangal had to face the risk of spurious seed to a higher extent than any non-cotton growing district farmer. To some extent, this also worked to the favour of debt-trapped farmers in the district, as the state policies, pressures from the political parties, presence of strong NGOs working against BT cotton, and media-hype made the purveyors of credit to restrain themselves from putting pressure on the farmers to repay debt in the Warangal district, in contrast to other districts.

Natural disaster related risks are often faced in the drylands. Pest risks are highly prevalent in districts which have mono-cropping and high percentage of area under non-food crops (Warangal, Guntur and Anantapur). Also, in these districts output risks, due to non-remunerative prices of cotton and groundnut constitute a major problem.

Based on the above discussion, review of literature and probit analysis (Galab *et al.*, 2008), the stylized facts in relation to farmers' suicides can be summarized as:

- A larger proportion of credit from formal institutions to the farmers reduces the probability of incidence of suicides.

- Expenditure on social events has a positive effect on the probability of suicides, which compounds the burden of indebtedness on the farmers.
- A larger investment on groundwater infrastructure reduces the probability of committing suicides provided the investment results in the expansion
- of irrigated area under groundwater sources and consequently, area under remunerative non-food crops is raised. However, the failure of non-food crops to yield returns increases the probability of suicides.
- There is no relationship between farm-size and incidence of farmers suicides.
- The more is the volatility of crop input and output prices, the higher is the probability of farmers' suicides
- The farmers whose household income is more diversified, have less possibility of committing suicides. But, if the income from non-agriculture is invested (current and capital investment) in crop production and the crop fails, the debt burden increases and chances of suicides also increase.

Responses to Agrarian Crisis

The study has revealed that due to lack of institutions (extension, credit, markets), transaction costs are too high in accessing inputs. The state has responded by announcing plethora of initiatives both, short-term and long-term. The spread of contract/corporate farming to large area and number of crops has been the response from the market. Similarly, there are responses from the civil society, distressed households and farming community.

State Response

Government of Andhra Pradesh: The Government of Andhra Pradesh has taken up some immediate (short-term) measures for mitigation of distress of the suicide-affected families.

Relief Package: Relief package to the brieved families in the form of ex-gratia amount of Rs 1 lakh and Rs 50,000 towards liquidation of farm debt and help lines are some of the initiatives in this regard. State initiatives have also come forth to stabilize agriculture, in the form of free power to agricultural connections, moratorium on loans for 6 months on private money lenders, two-year moratorium on institutional credit recovery by commercial banks, declared by Government of India, rescheduling of the bank loans of 7.93 lakh account holders, conversion to term loans, formulation of village credit plans (from *kharif* 2005), increase in the flow of institutional credit, stamp duty exemption to reduce the costs of borrowing for small farmers, and exemption of registration fees and stamp duty on loans up to Rs 1 lakh for small and marginal farmers,

Response related to Inputs and Insurance: These include crop insurance[3], approval of new State Seed Regulation Bill 2004 to regulate production and sale of seeds[4] and a comprehensive Bio-fertilizer Bill[5] to induce eco-friendly fertilizer-usage. The Seed Village Programme has been launched to enable the selected farmers to produce quality certified seeds. The foundation seeds are procured from the agricultural university for further multiplication. A compact area of 10 hectares covered by 50 farmers has been selected for the Seed Village Programme. The foundation seed is supplied on a subsidised basis.

Irrigation as a Priority: The state government has also taken up irrigation development programme in a big way with an estimated cost of Rs 93,000 crore. It is also mandated that lion's share of these allocations should be invested in less-irrigated regions of Telanagana and Rayalaseema. However, most of the ongoing projects would benefit the already developed coastal region. Even after successful completion of *Jalayagnam*, the ratio of irrigation between Telangana and Coastal Andhra would be at 1:2.7 (Rao, 2006). The regions facing acute agrarian distress which is manifested with high number of suicides were to get first priority in creation of public irrigation, but it is being denying to them.

Interventions in Input and Output Market: The state government has launched *Indira Kranti Patham* (IKP), a statewide poverty reduction project, to enable the rural poor to improve their livelihoods and quality of life through their own organizations. It aims to cover all the rural poor households in the state with a special focus on the 30 lakh poorest of the poor households. It is being implemented by the Society for Elimination of Rural Poverty (SERP), Department of Rural Development, Govt of Andhra Pradesh.

The IKP intervention of the state government has made some modest attempts to supply inputs to the SHG-members in the survey districts of Anantapur, Chittoor, Mahbubnagar and Srikakulam and Vizianagaram. For instance, in the Kotrapally and Gadduru villages of Chittoor district, some members of 14 SHGs had procured groundnut seeds at the rate of Rs 1800 per quintal, whereas the market price was in the range Rs 2200-2300/q. In Savara Jeribhadra, one of the survey villages in Srikakulam, the scheduled tribe members of SHG were made to purchase fertilizers collectively, reducing their dependence on middlemen and thereby, getting the benefits of lower transaction costs and increase in profits. Recently, the SHGs have taken a decision to formalize the informal debts to avoid the high interest burden on their members.

Market Intervention (Lac) in Adilabad

The SHG members of Pittabongaram village in Indravalli mandal of Adilabad district were guided by the 'DPIP Officials' to take the initiative in collecting the 'lac' (an important NTFP) and 'market' it themselves, as the price difference was vast when compared to the actual prices they got from local traders. Moreover, they were cheated in several ways (in terms of weight, grading, pricing etc.). Following their a proposal was drafted by CIG (production, collection and marketing) Pittabongaram (Jai Hanuman Sangham) to TPMU, Utnoor.

The objectives the CIG were to get remunerative prices for 'lac' and make its collection hassle free by entering into agreement with the forest officials. It was also intended to cover 25 villages surrounding Pittabongaram and encourage 'lac' growing on private lands by leasing-in land (with their CIG members). In the beginning, only 4 SHGs mustered courage to venture into this new means of livelihood. There were 76 members in all before forming a CIG for this purpose. In a proposal to TPMU, the CIG had asked for Rs 6,40,000, but an amount of Rs 2,50,000 was ultimately sanctioned. To oversee the affairs, a 'purchasing committee' of 12 members was constituted drawing 3 members each from every SHG. Two of these committee members were to manage the procurement of 'lac' on all days of the week.

To expand and acquire knowledge and skills on growing of 'lac' on private lands, 5 youths were sent for one week training to Indian Lac Research Institute at Ranchi in Jharkahand. On their return, the CIG leased-in 5 acres of land for 3 years at 12,000/annum and planted 8,000 saplings, which will take time to mature and yield profits. In the meantime, CIG members were able to purchase 43 quintals of 'lac' @ Rs 35-40/kg and sold the same at Rs 50/kg, after ascertaining the market price at Gondwa in Madhya Pradesh.

The IKP interventions are also being made in doing value addition in products like lac production, soya bean, cotton and red gram in

Adilabad; NTF and cashew in Srikakulam; red gram in Vizianagaram; milk marketing in Anantapur; red gram and castor in Mahaboobnagar and NTFP in Chittoor, by providing the required skills and knowledge. Interaction and exposure visits were also arranged for the youth.

A new system of decentralized purchase operations was started under the programme to save the farmers from unfavourable consequences of interlinked markets. The significant aspect of this scheme is the management by the Village Organizations (VOs) created out of the federation of the Self-help Groups (SHGs). The primary survey in Mahaboobnagar, Warangal and Karimnagar districts showed that VO procurement protected the marginal and small farmers from the retailer to a large extent. The sale to retailer and VO was transacted in the close neighbourhood of the seller to reduce the marketing costs. Earlier, when a farmer sold the crop to the retailer, the cost came to be Rs 5.20/q, while it was Rs 1.60/q when sold to the VO. The gain by selling to VO rather than to the retail trader was Rs 3423 per household; the gain for the marginal farmer was Rs 2455 per household. The gain for large farmers was twice that of the marginal farmers, indicating that the scheme was characterized by pro-poor bias.

Technology Interventions to Reduce Cost of Production: The indiscriminate use of pesticide and/or its failure to control pest attack, is often the source of high production cost/loss. Therefore, technology intervention through pest management programmes was considered vital for diffusing farmers' distress. Thus, the Non-pesticidal Management (NPM) model was scaled up in 11 districts in Andhra Pradesh as a state-NGO collaborative venture.

Institutional Interventions: In June 2003, the state government launched a programme for the formation of 'Rythu Mitra Groups' (RMGs), consisting of small, marginal and tenant farmers, share croppers and oral lessees in the state. Initially, each RMG was provided with Rs 2,500 for capacity building, book keeping and exposure visits. Each group is provided a loan of Rs 1 lakh, at an interest rate of 3 per cent, which the members would lend to the members for investment purposes. The RMGs, with about 15 members in each group, are expected to save and function on the lines of SHGs. The basic objectives of the RMG, in addition to credit, are technology transfer, and providing access to market information and market services to farmers. It is also envisaged that group approach would facilitate collective action in both input purchase and output marketing for mutual benefit of the members of a RMG and make the entire process participatory. It is also proposed by the state government to link all the departmental schemes like 'Seed Village Programme', 'ISOPAM', 'Farm Mechanization', 'Training Programmes', 'Vermi Compost', 'Green Manuring', etc. through RMGs so as to empower them for their sustainability.

Besides formation of RMGs, the state also attempted to empower the farmers by conducting 'Rythu Chaitanya Yatras' to educate farmers on the latest technology in agriculture and allied fields. More than 22 lakh farmers have been trained under this scheme. Ten allied departments participated in Rythu Chaitanya Yatra. The group of officers of agriculture and allied departments at the mandal level visited 54702 hamlets and conducted programmes of providing training to the farmers on technology gaps and simultaneously educated them on various government programmes and schemes. Focus was to increase the net income of farmers by reducing cost of cultivation. *Rythu Sadassus (Agricultural Exhibitions)* were organized at the divisional level, in May 2006 to focus on the strength of agriculture and allied fields, so that farmers could take full advantage of the exhibitions, lecturers, interactions and demonstrations.

Union Government: The Government of India (Prime Minister's package) has sanctioned an amount of Rs 9,650.55 crore towards rehabilitation of farmers in 16 selected districts of A.P. in order to curtail the suicide menace by distressed farmers. The schemes covered under Department of Agriculture mostly relate to water utilization, seeds and extension. The usage of credit component in the state has exceeded the target in case of interest rate waiver, but the credit needs also have not been properly assessed and provided for; hence, dependence on non-formal sources is still high. The non-credit component suffers from the problems of extensive coverage (watersheds) not so useful to the resource-poor farmers. Emphasis should be given to watershed development plus strategy should be to take the benefits to the needy.

Response of Civil Society Organizations

Provision of Extension Services: A IT-based personalized agro-advisory system, called 'e-Sagu', has been initiated for providing timely extension services to farmers. The main objective of 'e-sagu[6]' is to bridge the information gap by exploiting developments in information and communication technology. It aims to improve farm productivity by delivering high quality personalized (farm-specific) agro-expert advice in a timely manner to each farm at the farmer's door-steps. The advice/guidance is provided on a regular basis (typically once a week) from sowing to harvesting, which reduces the cost of cultivation and increases the farm productivity as well as quality of agri-commodities. In e-Sagu, the developments in IT (such as database, Internet, and digital photography) have been extended to improve the performance of agricultural extension services. The e-Sagu system offers next generation agro-advisory tools, and supplements and integrates the existing agricultural extension system.

The impact analysis has shown that farmers are satisfied with the expert advice as it is helping them to improve input efficiency by encouraging integrated pest management (IPM) methods, and judicious use of pesticides and fertilizers by avoiding their indiscriminate usage. The additional benefit accrued to farmers has been estimated to be Rs 3,874/-per acre, with cost-benefit ratio of 1:4. The turnaround time for advice delivery is 24-36 hours. At some centres, farmers have paid the nominal subscription fee for the services and they are satisfied with it.

Institutional Interventions: Although the farmers collectives/ cooperatives exist since long, NGOs have made fresh attempt in this direction in response to the collective felt-needs of the disempowered farmers. One of the NGOs in the Warangal district has promoted a few farmers cooperatives based on the philosophy of Cooperative Development Foundation (CDF). These cooperatives[7], though in the nascent stage, could supply inputs to its members at a lower price than the earlier arrangements of purchase of inputs (retail/wholesale trader). The case of a typical cooperative society shows five varieties of crops, viz. paddy, maize, cotton, chillis and other crops, for which saving due to price differences in inputs (seeds, fertilizer and pesticides) per acre would be Rs 110, Rs 150, Rs 320, Rs 350 and Rs 95, respectively. The total gain to the farmers per acre for all the crops put together would amount to Rs 248 on paddy, Rs 337 on maize, Rs 720 on cotton, Rs 787 on chillis and Rs 213 on other crops, which shows that the gain is the highest for non-food crops of cotton and chillis[8]. On the aggregate, on all crops the benefit to the farmer is Rs 2312 per acre[9].

Thus, the farmers' cooperatives[10] are able to save on the input costs, reducing the cost of production and the benefit is high in the case of cash crops like cotton and chillis. This proves the huge potential the cooperatives possess on working collectively on reducing cost of production, and getting right price by not resorting to distress sales. This model can be replicated and assisted by the government by either promoting farmers' cooperatives/self-help groups by itself or through the NGOs.

Response of Market

e-Safal[11]*:* A private initiative with the help of *Krishi Vignan Kendram*, an agricultural resource centre and ICAR, integrates farmer's production decisions, farm management, extension and marketing, by linking up the agencies that supply services in five villages of Jammikunta mandal in the district of Karimnagar in A.P. After conducting a few educational camps in each of a dozen villages with power point presentations combined with visual aids to inform and educate farmers about the internet-linked state-of-the-art technological facilities and possibilities of new farm management, farmers who volunteered to come forward to join the programme were enlisted on a nominal fee of Rs 100. The project liaisons between the farmers' clusters, local *Krishi Vignan Kendra*, JK Seeds, Synzenta (a pesticides supplier) and a super-spinning mill from Coimbatore. Farmers were provided with free soil testing facility with the KVK and the reports were sent to JK Seeds. A detailed database of all the farmers was generated with a website at KVK, which can be accessed through the net. The JK Seeds examined the soil reports and suggested the suitable seeds. JK and Synzenta are given a captive market. Both the companies fund extension workers stationed at headquarters and provide weekly free consultations. Farmers are also advised on integrated pest management techniques. Field staff also educates them about the precautions to be taken at the farm level such as separation of cotton according to quality and prevention of contamination. Finally, the Super Spinning Mill is contracted to purchase the cotton at a premium price. All the efforts, finally lead to a substantial reduction in cost of production, increased yield, and quality and value for the product.

Contract/Corporate farming: Contract farming is not new and can be dated back to 19th century with respect to crops like tobacco, cotton, indigo. In the current context, it is considered as a panacea to the agricultural deceleration with the potential to bail out the farming community from the unprecedented spate of suicides. The experience of 'Large Scale Advances Farm Project', popularly known 'Kuppam project' in contract/corporate farming for the past few years in AP is a case in point to examine the effects of contract farming. At only a small fraction of supporting the corporate farming by the state, peasant farming can be made viable and vibrant absorbing all modern technological development and meeting the needs of future (Chowdry *et al.*, 2002).

Response of Distressed Families[12]

The consequent effects of suicides have remained less explored compared to the causes that led to suicides. The response from the deceased farmers' families in the post-suicide situation highlights, one, the organization of agriculture by them and two, the other side of the state response to the ultimate form of distress or suicide. Few case studies were undertaken in depth to understand these issues from the sample districts of Warangal, Anantapur, Mahbubnagar and Guntur.

The pressure for repayment of debt was so high even after the suicide that Vijaya from Issipeta (Warangal district) had to put all her land to sale.

> *"After the death of Sudhakar Reddy I could not face the pressure of the debtors. Then I sold off the whole of 2.5 acres. Actually, the land was divided among the debtors, fertilizer dealer took one acre under debt, three other debtors took one acre, 20 guntas (half acre) was given to debtors from my natal village. They tried to sell the land but nobody purchased it. There was a total debt of 2.5 lakh. The Sarpanch and village elders made a settlement regarding the debts. Still I have to repay debts to my sisters"* *(Yara Vijaya).*

There were mixed responses with regard to organization of agriculture and shift in livelihoods. Women workers who were engaged as hired or family labour prior to commitment of suicide by their respective husbands, had to shoulder the entire responsibility of cultivating land or diversify from agriculture labour to casual wage work to support the family. Women belonging to the middle and lower social groups settled most of the debts with the help of government relief package, leased-out land and took to wage labour, thereby slipping into the unstable occupations and incomes. In general, there was an increase in the work burden for the women in the post-suicide situation.

The state has declared the compensation package for the suicide victims as a policy in 2004. Though suicides took place earlier from 1997 onwards, the then government did not identify it as a social calamity[13]. The relief package came after a lag of 7-8 years either freshly or in the form of 5-year bonds[14]. However, the award of compensation was wrought with problems, where the opportunity and transaction costs were high.

> *'As I cannot live alone with a young child, my brother-in-law stays with us, although he is not of much help. But his presence is a big solace. He went round the mandal office for the sanction of government grant. Sometimes he abuses and I bear all that. I felt low when my sister-in-law quarrelled for her share in the land that we had been cultivating'* *(Nirmala).*

> *'Let the* Sarkar *(government) be blessed, it gave us Rs 1,00,000 in 1998. They helped with a cash ex-gratia of Rs 25,000 and Rs 75,000 as Indira Vikas Patra bonds locked in for five years. Bonds worth Rs 25,000 were put on my name and worth Rs 50,000 were put on the name of my daughter, which matured to Rs 1.5 lakh by 2003. I performed the marriage of my daughter in the same year and gave Rs 1.2 lakh as dowry to my daughter. Part of the amount I received (Rs 50,000) also I gave as dowry (Rs 20,000) and the rest I spent on marriage expenses. My sons asked for a share in this but I said that the girl's marriage is important. How I could have performed the marriage if the sons had taken away the money? Of course, my sons did not oppose me on this"* *(Ch Sammakka).*

> *'I received Rs 25,000 in cash from the mandal revenue officer, this I paid it off for the adtidar in Warangal. My brothers and brothers-in-law looked into the settlement issues and also dealt with the officials. We also incurred some expenses for accessing the relief package. The rest of Rs. 75, 000 was put as fixed deposit on my name. As it matured to Rs 1.5 lakh last year (2004), I*

used it to pay off part of the promised dowry amount to my daughter, some amount of debts were also repaid (Hymavathy Reddy).

Policy denying assistance to those affected by it in the name of 'rules' can not be viewed as a lesser curse than the death that is befallen on those poor widows. Government assistance is dismal as it fails to look the situation from the gender sensitivity. Technicalities like a cut-off date for the award of compensation denied the same in the case of Yara Vijaya.

Women farmers and farmers who have no land titles have been denied the status of a 'farmer' and hence compensation package.

Conclusions

Agrarian distress is pervading all around the state, but the causes and consequences are different, depending on the variability of development levels within the regions. In developed pockets though resource endowment is not a problem, the yields of major crops (growth) have tapered off at a low level with comparable south Indian states. The transformation in land ownership has resulted in a large proportion of agricultural labour, thereby increasing tenancy. The development of land under tenancy is hampered due to the oral nature of tenancy contract, which precludes any investment on the side of lessees and leasers.

Another critical issue is the falling area under canal irrigation. The new institutions carved to mitigate the problems thereof also have not yielded desirable results. Therefore, participatory management institutions need to be given power in the real sense of decentralization. In the case of drylands, a major shift towards high-value non-food crops has resulted in depletion of water resources and ecological degradation. Though shift in cropping pattern has resulted in economic growth in the short-run, in the long-term, the sustainability of growth would be problematic. Though major irrigation projects have been taken up by the present political regime, it is doubtful whether they would fulfill equity considerations which is very important to such regions. In this context, the World Bank project on rejuvenation of tanks is a welcome move.

The technology stagnation is being overcome in the present times with newer initiatives, like biotechnology and genetically modified technology. As the relative benefits of these technologies are still debatable, their dissemination to the farmers for a wider application remains limited. The market response to new technologies is limited to few high-value crops. Hence, state needs to take up a more rigid regulatory role and itself give more priority to such research. Also, the programmes taken up to address the issues of agricultural extension need to be more institutionalized instead of a makeshift approach.

Output market volatility is a reality and to minimize its impact, adequate infrastructure needs to be created. More investment on storage, roads and communication is wanting. Alternative experiments involving public-private partnerships in marketing (IKP), lowered input costs (NPM/IPM), and technology—based extension system (e-sagu), are a ray of hope in this context. This public-private partnership has facilitated far wider spread and dissemination of novel experiments which is a welcome trend in the state.

On the other hand there is a felt need for organizing the farmers into collectives/cooperatives so as to negotiate with the changing market conditions. In the age of multiplicity of market arrangements, cooperative coordination based on the principle 'service at a cost' alone can give voice to the people and empower them. The onus of promoting such initiatives rests with the state.

Notes

1. This paper is a part of larger on going project on 'Agrarian Distress in Andhra Pradesh', CESS, Hyderabad.
2. This Section is Largely based on two papers submitted by the authros to the book on 'Agraian Distress in India', IGIDR.
3. Restoration of 50 per cent of premium subsidy to small and marginal farmers, enhancement of indemnity levels to 80 per cent, reduction of premium rates to 2 per cent for cereal crops and 3 per cent for commercial crops and payment compensation in two spells for *kharif* and *rabi* crops.
4. Though the state government has approved the New Seed Regulation Bill, but it has not been made into law.
5. The Bio-fertilizer Bill has been passed in the Andhra Pradesh Legislative Assembly in 2006
6. Sagu in Telugu means cultivation
7. The number of cooperatives formed by the NGO is ten (in villages belonging to the mandals of Issipet, Mogullapally, Dharmasagar, Hanamakonda, Wardhannapeta, Shayampeta and Sangem) with a

membership of 675 and a share capital of Rs 52,000. The total deposits are Rs 10,91,000 and the total funds stand at Rs 21,38,995. The value of total purchase of inputs (seed, fertilizer and pesticides) by the societies is Rs 50,05,215 and the value of total sales is Rs 45,41,386. This NGO has not accessed any funds from government and all of them are registered under the MACS Act.

8. The total gain (per acre) includes price difference + bonus + transaction costs which would have incurred had the farmer purchased inputs from the open market.
9. This aggregate benefit will be enjoyed by the farmer only if he cultivates all the five crops.
10. These cooperatives have not yet started marketing of produce but which will be started from this *kharif* onwards.
11. e-Safal is an initiative started by the Satyam Computers group in 2005. The experience of this initiative is documented by the project on 'Vertical Integration in Cotton Farm ing in AP: A Study of e-Safal' (interim report by R.V. Ramana Murthy, University of Hyderabad, 2007)
12. This section is based on the project report on 'Qualitative study of farmers' suicides in Andhra Pradesh', carried out by the authors.
13. A few cases were awarded compensation as ex-gratia but not as a policy by the Telugu Desam Govt.
14. Relief packages are of two types: (a) Rs 1 lakh at the time of suicide during 1997-98 that was split as Rs 25,000 cash as ex gratia, and (b) Rs 75,000 Indira Vikas Patra bonds of 5 years in the name of the woman and girl child.

References

Rao, C.N. (2006), Agrarian crisis in Andhra Pradesh, *Journal of Indian School of Political Economy*, 18(1&2).

Chowdry, K.R., A. Prasada Rao, Venkat and Uma Shankar (2002), *AP Coalition in Defence of Diversity and Deccan Development Society*, 'A Report on Kuppam project'.

Galab, S. and E. Revathi (2006) *Changing Agrarian Relations and Farmers Collective Actions in Dryland Agriculture: A Perspective*, Unpublished paper, CESS

Galab, S. and P. Prudhvikar Reddy (2007), Agriculture crisis in Andhra Pradesh: Evidence from the field, Paper submitted to the Book *Project on Agrarian Crisis in India*, IGIDR, Mumbai.

Galab, S., P. Prudhvikar Reddy and R. Hinaz (2008), Young lives; Round 2 Survey Report—Initial findings, Andhra Pradesh available at www.younglives.org.uk/publications/country_reports

NSSO (National Sample Survey Organisation) (2005), *Situation Assessment Survey of Farmers: Indebtedness of Farmer Households*, NSSO 59th

Round, Report 498. NSSO, Ministry of Statistics and Programme Implementation, Government of India, New Delhi.

Reddy, Narsimha D. (2007), *Farmers' Indebtedness and Agricultural Credit Situation in Andhra Pradesh*, Paper submitted to the Committee on Indebtedness of Farmers, Government of India.

Reddy, K.S. and P. Prudhvikar Reddy (2007), Banking sector reforms and equity: A sub-regional analysis, In:–Chapter 30, *Fifty Years of Andhra Pradesh 1956–2006*, Eds: R.S. Rao, V.Hanumantha Rao and N. Venugopal, Centre for Documentation, Research and Communication, Hyderabad.

Revathi, E. (2006), *Suicide as Vulnerability: Some Dimensions of Agrarian Distress in Andhra Pradesh*, Centre for Economics and Social Studies, Hyderabad.

Revathi, E. (2007), Farmers' suicides in Andhra Pradesh: Issues and policy concerns Paper submitted to the Book Project on '*Agrarian Crisis in India*', IGIDR, Mumbai

Subramanyam, S. and P. Aparna (2007), *Growth and Structural Changes in Andhra Pradesh Agriculture*, Background paper submitted to Human Development Report, CESS, Hyderabad.

Section III

OPPORTUNITIES FOR ALLEVIATING AGRARIAN DISTRESS

The current situation of agrarian distress is undoubtedly serious although the drivers of demand offer opportunities for agricultural growth. Population growth, urbanization and income growth in developing countries are fuelling a massive global increase in demand for high-value agricultural commodities like fruits, vegetables and livestock products. The demand projections reveal that in the South Asian countries, the consumption of fruits, vegetables, meat, eggs and fish will increase by more than 90 percent in next 15 years of so. The demand for low *fat* based value-added milk products is also rising rapidly in India and other Asian countries. The shift in the food basket away from traditional staple foods due to changing consumer preference is creating new opportunities for the farmers to diversity cultivation to high-value commodities (HVCs).

The dominance of smallholders in the Indian agricultural is often considered to binder the ' diversification of agricultural production towards HVCs as the smallholders have low marketable surpluses and weaker resource position *to* effectively integrate themselves into the market-led growth. However, the milk cooperatives in Gujarat, grape growers' association in Maharashtra, and several other such institutional innovations have demonstrated the success oflinking small producers to national and international markets.

Section III looks into the emerging opportunities provided by buoyant demand of high-value agricultural commodities for the smallholders to come out of the low-return trap in agricultural activity. The challenges that smallholders will have to cope with and the enabling institutional and policy requirements for facilitating them in harnessing the opportunities are discussed in detail. Case studies covering various institutional arrangements such as Cooperatives, Growers' Association, Contract Fanning and Retail Chains arc presented- to understand the dynamics of smallholders, participation in production of high value commodities.

Besides agricultural diversification, development of rural *non-farm* sector offers opportunities for mitigation of agrarian distress. Construction, trade, transport and business services have spear-headed the non-farm employment growth in the rural sector. Although these industries together account for only 11 per cent of rural workforce in the country, the emphasis that rural infrastructure development has received in the recent years has the potential to transform these industries as the main drivers of rural non-farm employment growth and a pathway out of agrarian distress in India. The study on rural uon-rarm employment examines the extent to which employment growth in these sectors can drive rural diversification in India and ascertain how growth in rural livelihood can be sustained.

8

Role of Institutional Innovations in Linking Farmers to Markets in the Changing Agricultural Scenario

Pratap S. Birthal and P. Parthasarathy Rao

Market liberalization and globalization are causing a transformation in agriculture and agri-food markets in India. Increasing income and changing preferences are leading to a shift in the food basket away from the traditional staple foods to high-value products like fruits, vegetables and animal products (Joshi *et al.*, 2003; 2004). The increased demand for high-value commodities (HVCs) is creating opportunities as well as challenges for the farmers. There are strong monetary incentives for the farmers to diversify cultivation to high-value commodities due to their strong potential for higher returns to land, labour and capital. However, dominance of smallholders in Indian agriculture is a potent stumbling block. On one hand, while the small farm-size makes it relatively easier to shift from the cultivation of traditional crops to that of high-value commodities owing to lower overheads, on the other hand, the small size implies that the marketable surpluses are small, leading to high transactions costs. Also, it is feared that smallholders will not be able to meet the quality standards required by the industry owing to lack of access to the right technology and infrastructure. Thus, there are apprehensions that processors (fruit, vegetable, milk, meat, feed, alcohol, etc.) and super market chains would overlook the small farmers since they need bulk quantities of specified qualities year round.

This Chapter has examined the opportunities and challenges for the smallholders in market-oriented production of high-value commodities and has identified enabling institutional and policy requirements for their participation.

Opportunities in High-value Agriculture

The trends show that for rural consumers, the share of high-value food commodities (fruits, vegetables, animal products and beverages) in total food expenditure increased from 30.4 per cent in 1983 to 44.1 per cent in 2004-05, while the share of food grains declined from 55.3 per cent to 38.6 per cent (Table 8.1). Urban consumers spend relatively more, 55.4 per cent of their food expenditure, on high-value foods. Nonetheless, consumption of high-value commodities in rural areas has been growing faster, indicating a tendency of convergence in the rural-urban disparities in their consumption. At the same time, with growing demand for livestock products, the derived demand for coarse cereals, soymeal, etc. is expected to rise. For example, the poultry sector is growing at the rate of 8-10 per cent and since coarse cereals constitute up to 50 per cent of the poultry rations, their demand by animal feed industry is increasing (Klieh *et al.*, 2002). The policy on mandatory blending of ethanol with petrol up to 5 per cent and increasing it to 10 per cent by the end of 2017 is expected to further boost the demand for coarse cereals in the ethanol industry.

Table 8.1: Shares of Different Food Items in Total Food Expenditure in India *(in per cent)*

Food items	Rural		Urban	
	1983	*2004-05*	*1983*	*2004-05*
Cereals	49.5	32.8	32.9	23.8
Pulses	5.8	5.8	5.8	5.3
Milk and milk products	11.5	15.4	15.7	18.6
Edible oil	6.2	8.4	8.2	8.1
Meat, egg and fish	4.6	6.0	6.1	6.4
Vegetables	7.2	11.1	8.4	10.5
Fruits	2.1	3.4	3.6	5.3
Sugar	4.3	4.3	4.2	3.5
Salt and spices	3.8	4.5	3.6	3.9
Beverages	5.0	8.2	11.6	14.6
Share of food in total expenditure	65.6	55.0	58.7	42.5
Total	100	100	100	100

Source: GOI (2006a)

Changes in consumption pattern are being driven by sustained income growth and urbanization. Between 1980-81 and 2004-05, per capita income in India grew at about 4 per cent a year, and urban population about 3 per cent a year. Expenditure elasticity for high-value food products ranges from 0.80 to 1.04, much higher than for any other food item (Ravi and Roy, 2006). Besides, changing lifestyles, increasing entry of women in workforce, growing nuclear families, increasing use of credit cards, improvements in transport infrastructure and rise of supermarkets have also facilitated these changes. Ravi and Roy (2006) have predicted demand for high-value food products to grow at an annual rate of around 5 per cent by 2020, as against 2.5 per cent for food grains.

Globalization is creating opportunities for the export of high-value products. Evidence indicates an accelerated flow of exports of high-value food products from developing to developed countries (Diaz-Bonilla and Recca, 2000; Aksoy, 2005). Though India is not a significant player in the global trade in agricultural products, it has considerable share (10%) in exports of commodities, like mangoes and onions. Horticulture products account for 16 per cent of total value of exports next to rice. Meat and meat products accounted for 5.3 per cent and their share has increased substantially over time.

High-value Agriculture and Smallholders

In response to the growing demand, growth in high-value segment of agriculture accelerated from 4.2 per cent during 1980s to 4.8 per cent during 1990-91 to 2003-04 (Table 8.2). High-value food commodities now account for about 47 percent of the value of agricultural output, up from 35 per cent in TE (triennium ending) 1982-83. Changes in the production portfolio are starker at the commodity or commodity group level. During this period, share of fruits and vegetables increased from 14 to 17 per cent, milk from 13 to 18 per cent, poultry from 1.6 to 3.2 per cent, and fish from 2.6 to 4.3 per cent. This provided a cushion to the agricultural growth, which otherwise would have decelerated at a higher rate.

The robust growth in high-value agriculture was facilitated by policies such as establishment of the National Horticulture Board in 1984 to promote the production, processing, marketing and exports of horticultural products; and easing out the Milk and Milk Products Order (MMPO) that restricted private sector participation in the

Table 8.2: Changes in the Composition of Agricultural Sector

Items	Share in value of output, %		Annual compound growth, %	
	TE 1982-83	TE 2003-04	1980-81 to 1990-91	1991-92 to 2003-04
Crops	77.3	69.3	2.6	2.2
Cereals	27.4	22.9	3.0	1.3
Rice	14.6	12.5	3.6	1.0
Wheat	7.9	7.5	3.4	2.1
Pulses	5.6	3.4	1.2	-0.2
Oilseeds	6.6	7.1	5.3	0.8
Fibres	3.4	2.7	5.3	0.8
Drugs and narcotics	1.6	1.8	1.8	5.0
Condiments and spices	2.1	2.5	4.6	3.9
Fruits and vegetables	14.0	17.3	2.7	5.4
Sugars	6.0	5.6	3.2	2.3
Livestock	20.0	26.4	4.7	4.1
Dairy	12.9	17.5	5.5	4.1
Poultry	1.6	3.2	6.2	6.3
Meat	1.4	1.8	5.5	2.9
Fisheries	2.6	4.3	3.2	5.9
Agricultural sector	100.0	100.0	3.1	2.8
High-value foods	34.6	46.6	4.2	4.8
Rest of agriculture	65.4	53.4	2.5	1.4

Source: GOI (various years)

dairy industry. Beside these, the government also provided fiscal incentives, such as reduction in corporate taxes and excise duties on the processed foods as a part of the economic reforms programme initiated in 1991. De-licensing of food processing industry has also attracted considerable foreign direct investments (FDIs), leading to a faster growth in processed food production.

Indian agriculture is dominated by smallholdings (< 2 ha). The share of smallholdings in the total number of holdings has steadily increased from 75 per cent in 1981-82 to 86 per cent in 2002-03 (GOI, 2006 b). High-value agriculture is suited to smallholder production systems as production of high-value agricultural commodities is labour intensive and smallholders have sufficient family labour. Contrary to expectations, they have a larger presence in high-value segment. Over 84 per cent vegetable growers and 88 per cent fruit growers are smallholders (Table 8.3). They account for 61 per cent of the vegetable area and 52 per cent the fruit area. Their presence in milk production is also considerable; over 77 per cent milk producers belong to smallholders and contribute 69 per cent to total milk production (GOI 1999; 2005).

Table 8.3: Participation of Smallholders in Production of High-Value Commodities

Category	*Vegetables*	*Fruits*	*Diary*
Participation rate (%)			
Small (? 2.0ha)	15.8	5.0	41.0
Medium (2.0–4ha)	14.8	2.7	56.7
Large (?4.0 ha)	10.4	3.0	68.5
All	15.3	4.6	44.2
Distribution of participating households (%)			
Small (? 2.0ha)	83.5	88.4	77.4
Medium (2.0–4ha)	11.9	7.1	13.7
Large (?4.0 ha)	4.6	4.5	8.9
All	100.0	100.0	100.0
Share in area/production (%)			
Small (? 2.0ha)	61.0	51.9	68.8
Medium (2.0–4ha)	20.9	19.3	16.7
Large (?4.0 ha)	18.1	28.8	14.5
All	100.0	100.0	100.0

Source: For fruits and vegetables, GOI (1999); for dairy GOI (2005).

However, the issue is how smallholders can participate in high-value agriculture and meet the requirements for the industrial/processing sector, and supermarkets who need bulk quantities of pre-specified quality and timely delivery. Similarly, in the context of the opportunity offered by the globalization of agri-food markets, the important issue is to create an enabling environment for the Indian small farmers to compete in the global market and reap the benefits of participatation in global food supply chains by increasing exports. Although, producer prices of commodities like milk, pig and bovine meat, eggs, bananas and grapes in India are lower than in many other countries, indicating its comparative advantage in production (FAOSTAT, 2007), India is at a disadvantage in the international market due to lack of scale economies in production and processing and high cost of compliance of the stringent food safety and quality standards in the global market. It is generally felt that small holders are constrained by lack of capital, inputs, technology and services, and access to markets which may act as a barrier to their diversification towards high-value agriculture to meet the domestic and export requirements.

Improving Small Farmers' Access to Markets Through Institutional Innovations

Smallholders though make sizable contribution to high-value food production, their access to markets is constrained by scale. Their marketable surplus is small, while local markets for high-value commodities are thin and sale in distant urban markets raises transportation and marketing costs. Existing supply chains are long and are dominated by a number of intermediaries like assemblers, wholesalers, sub-wholesalers, commission agents and retailers. In the case of fruits and vegetables, farmers receive one-third to one-half of the final price (Gandhi and Namboodiri, 2002), indicating high marketing costs and margins. Birthal *et al.* (2005) have estimated marketing costs at around 20 per cent of the sale price of vegetables. Marketing costs are also high in the case of milk, 15-20 per cent (Birthal *et al.*, 2005; 2006).

Similarly, for coarse cereals traditionally small farmers have been growing these commodities for home consumption and marketing small surpluses in the food grain markets either directly or through brokers. Owing to small surpluses, the marketing and transaction costs are generally high (CFC-FAO-ICRISAST, 2006). However, under the changing agricultural scenario industrial processors are demanding bulk quantities of these grains. There is thus, a need to link small-scale growers to industry.

High marketing and transaction costs act as a barrier to farmers' participation in markets. Institutions such as cooperatives, growers' associations, bulk marketing, and contract farming are considered to reduce marketing and transaction costs and risks by providing markets to farmers at their doorsteps (Eaton and Shepherd, 2001). Some of the successful models of such institutions are:

Cooperatives: India's dairy cooperative model is one of the successful models for linking farmers to markets that saves the producers' from exploitation by the middlemen/informal traders, and improves their bargaining strength and economies of scale in marketing. In 2004-05, there were more than 11.3 thousand village level dairy cooperative societies, with 12.3 million members, procuring 7.3 million tonnes of milk, equivalent to 7.9 per cent of the total milk produced in the country (NDDB, 2005). Cooperatives also provide inputs, services and information to the producers.

Cooperatives also exist for other products including oilseeds, fruits and vegetables. Two important cooperatives in horticulture are: Mahagrapes in Maharashtra and HOPCOMS (Horticulture Produce Cooperative Marketing and Processing Society Limited) in Karnataka. Mahagrapes is an association of grape-growers' cooperative societies. Mahagrapes was established in 1991 to improve grape growers access to domestic and international markets, which otherwise was difficult for individual producers or cooperative societies. It now has 16 growers' cooperatives with 2500 farmer members. Mahagrapes supplies inputs, technology, and extension services to farmers through cooperatives and empowers them to produce quality output conforming to food safety and quality standards of the importing countries. Cooperatives associated with Mahagrapes have refrigerated transport and cold storage facilities, for which cooperatives charge a fixed amount, on per kg basis, from the farmers. In 2003-04, Mahagrapes exported grapes worth US$ 2.2 million, mainly to UK, Netherlands and Middle East countries. It charges a fixed amount from farmers, on per kg basis, to meet costs of transportation, labour and other activities. The profits are passed on to the farmers. The net revenue for farmer members was about 60 per cent higher than those selling in the open market (Bakshi *et al.*, 2006).

Established by the state government in 1959 in Bangalore, HOPCOMS now secures its supplies from about 12000 farmer-members through its procurement centres. It markets about 500 tonnes of vegetables daily through 504 retail outlets. It also has 5 cold storages and one processing unit. The Society also has outlets for supply of inputs to the farmers.

Growers' Associations: Growers' associations are informal cooperatives managed by farmers themselves. SAFAL—village level associations promoted by the Mother Dairy Fruits and Vegetables Limited (MDFVL)—has been quite successful in linking fruit and vegetable farmers to markets. SAFAL was first established in 1988 to cater to the growing demand for fruits and vegetables in the metropolitan city of Delhi. At present, there are 250 SAFAL associations with about 20000 farmer members in the country.

MDFVL, an ISO-9002 and HACCP certified firm, provides technical support to these associations in the preparation of crop calendar and sowing schedules to get the desired supply on a regular basis. The firm also provides inputs like quality seeds, bio-pesticides and bio-fertilizers, and extension services to the producer-members. Quality standards for each fruit and vegetable are defined in respect of size, weight, colour and appearance, and are displayed at each SAFAL collection centre for farmers to comply with.

Daily wholesale market prices in the Delhi wholesale market serve as a base price for the producers. Farmers are paid the modal price plus a 5-10 per cent premium for quality. The firm does not share any production and price risk.

Farmers are being benefited from SAFAL supply chain. A study on spinach production has shown that SAFAL farmers, on average, realized 78 per cent higher profits, 8 per cent higher prices and incurred 92 per cent less marketing costs over those supplying in the open market (Table 8.4).

Table 8.4: Economics of Contract and Independent Spinach Production in National Capital Region Delhi, 2001-02

Particulars	Contract	Independent	% difference
Yield (t/ha)	8.6	8.3	4.0
Production cost (Rs/t)	1485	1630	-8.9
Marketing cost (Rs/t)	35	437	-92
Total cost (Rs/t)	1520	2067	-26.5
Price (Rs/t)	3311	3074	7.7
Net revenue (Rs/t)	1791	1007	77.9

Source: Birthal *et al.* (2005)

There are also growers' associations for other commodities. Two notable examples include Self-Help Groups (SHGs) promoted by Appachi Cotton Company Limited in Tamil Nadu, and Agrocel Pure and Fair Cotton Growers' Associations promoted by Agrocel Industries Limited in Gujarat.

Contract Farming: Recent reforms in the agricultural markets have opened up new avenues for agribusiness in India. Many a big business houses have started entering into the agricultural markets, and use contract farming as a means to source their raw material requirements. Contract farming is practised in many agricultural commodities, including wheat, Basmati rice, fruits, vegetables and medicinal plants, but is prominent in poultry and dairy. About 40 per cent broiler production in the country is under contract. For milk, no such estimate is available. But private sector procures nearly 8-10 per cent of the total milk produced in the country.

The agribusiness firms follow different models of contract farming ranging from bi-partite to multi-partite agreements. A bi-partite contract is an agreement between producers and firms, which may take the form of a market specification or resource-providing contract. Inclusion of other agencies by the agribusiness firms to provide inputs, credit and insurance gives rise to multi-partite contracts.

In India bi-partite contracts are common in poultry. Firms provide critical inputs like day-old chicks, feed, medicines, vaccines and extension services. Poultry farmers bear the cost of labour, shed, litter, water and electricity.

Farmers are provided a guaranteed remuneration in the fixed growing charges based on bodyweight of the birds. The firms lift the entire output and thus, farmers are insulated against market risks. From the farmers' perspective, most important benefit of contract farming is their insulation against price risk that is very high in poultry production (Figure 8.1). Ramaswami *et al.* (2006) have shown that through contract farming farmers could shift as much as 88 per cent income risk to the firms. Another benefit is the access to interest-free credit in the form of inputs. In a study, Birthal *et al.* (2005) have estimated 58 per cent reduction in marketing and transaction costs and 13 per cent increase in net profits due to contract farming.

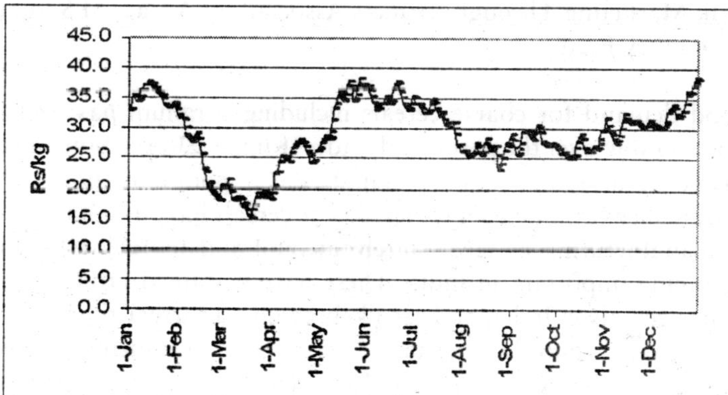

Figure 8.1: Seasonal Variation in Farm-Gate Price of Broiler: 2004-06

Tri-partite contracts are common in dairying. Firms enter in contract with a local person in the village who acts as an intermediary/ facilitator between producers and firm. This reduces transaction costs of contracting with a large number of small-scale producers. Bi-partite contracts are with large dairy producers.

The intermediary (agent) also assists firm staff in dissemination/ distribution of information, inputs, services and milk payments to producer-suppliers. Price of milk is determined on the basis of its fat and SNF (solids non fat) contents, taking into consideration the prices fixed by its competitors.

Main benefit from contract farming in milk is the reduction in marketing and transaction costs. Birthal *et al.* (2005) have shown that with contract farming, farmers could save as much as 93 per cent of these costs. This enabled them to earn almost twice the net revenue as compared to independent farmers (Table 8.5).

Table 8.5: Economics of Contract Milk Production in Punjab: 2001-02

Item	*Contract*	*Independent*	*Difference, %*
Milk yield (kg)	11.9	11.4	4.4
Production cost (Rs/t)	5586	5782	-2.5
Marketing cost (Rs/t)	100	1442	-93.1
Total cost (Rs/t)	5686	7170	-20.7
Price (Rs/t)	9337	8991	3.8
Net revenue (Rs/t)	3651	1821	100.5

Source: Birthal *et al.* (2005)

Bulk Marketing Through Farmers Association: A Case of Sorghum for Poultry Feed

Food demand for coarse cereals including sorghum has declined considerably over the past two decades (Kumar, 2006). Nonetheless, coarse cereals remain important in the diet of poor households in the marginal environments. Besides, coarse cereals have alternative uses in manufacturing as feed, starch, alcohol and biofuel, which are becoming important in India. These have sustained production of coarse cereals in India, and are likely to give a further fillip.

Resource-poor farmers in the marginal environments, cultivate mainly rainy-season sorghum in India. However, owing to scattered and small-scale production, farmers are unable to meet the requirements of the industry that need grain in bulk quantities. After harvest, the surplus sorghum grain is sold either in a regulated market through commission agents or through a broker (middleman) at the village. Despite several inefficiencies and exploitation by middlemen, Marshland and Parthasarathy Rao (1999) found that the marketing system *per se* did not constrain the utilization of the crop as food grain. At the same time, the study concluded that the existing marketing system was not optimal for industrial users, who would prefer to obtain sorghum through new institutional arrangements such as, contract farming and bulk purchasing that compress the marketing chain and thus, reduce transaction and marketing costs.

In order to bridge this gap, ICRISAT took an initiative on a pilot scale with the objective of linking small-scale sorghum producers with poultry feed manufacturers through informal institutional arrangements. In the process of linking sorghum producers and poultry feed manufacturers, many simultaneous activities were carried out. For instance, Poultry Feed Trials (PFT) to further corroborate the efficiency of sorghum in poultry rations, supply of improved sorghum seed and technology to farmers, formation of Farmers' Associations, training in bulk storage and bulk marketing. There were a number of advantages to both, the buyers and sellers due to bulk marketing. Farmers could have an easy access to market, reduce input costs due to backward linkages with input dealers, reduce marketing and transaction costs and improve their bargaining power. Buyers could have an assured supply quality, reduce marketing and transaction costs and enforce safety standards.

This model is now successfully operating in 5 clusters in the states of Maharashtra and Andhra Pradesh. For the success and sustainability of this model of bulk marketing through formation of farmers' associations, it was felt that there was a need for credit from formal sources since the linkage between the credit and output market, i.e. tied transactions was one of reasons for distress sales outside the Farmers Association. There is also a need to explore informal/formal agreements through novel market linkage models for the long-run sustainability of the market linkages between the farmer and the feed industry/processors. Although the transactions are based on mutual trust, a formal agreement would provide some kind of a cushion to both the parties. Finally, there is a need for a more critical examination of the roles and responsibilities of Farmers Association.

Participation of Smallholders in Institutional Innovations

Smallholders are expected to benefit most from institutional marketing arrangements, as their marketing and transaction costs are higher. However, their inclusion in the supply chain may be limited as agri-business firms tend to contract with those farmers who are capable of supplying large volumes and complying with their food safety and quality standards. Smallholders are often excluded from the institutional innovations on the ground that contracting with a large number of them raises transaction costs (search, monitoring and enforcement) to the firms.

There is a mixed evidence regarding participation of smallholders in the coordinated marketing systems. In contract farming of spinach (Birthal and Joshi, 2007) and of gherkin (Erappa, 2006) it was found that more than 50 per cent farmers were small landholders (= 2ha). Birthal *et al.* (2005) have reported considerable involvement of small-scale producers in contract farming of milk, but not in broiler production. In dairying, 56 per cent contract farmers had = 5 milk animals, while in broiler production, only about one-third of the contract producers had = 5000 birds/cycle. On the other hand, Kumar (2006) has found very little involvement (15%) of small landholders in crop production in Punjab.

It may, however, be mentioned that contracting with smallholders is advantageous to the firms. One of the major advantages is that contracting with a large number of smallholders reduces supply risks, which could be very high if the contracts are with a few large producers, and there are limited alternative supply sources. Second, it is easier for the firms to enforce contracts with smallholders because of their poor bargaining power. Third, though smallholders have tiny marketable surpluses, they are capable of producing at a lower cost because of sufficient availability of family labour. The sufficient availability of family labour also enables them to better comply with food quality and safety standards which include many a cultivation practices (manual weeding in place of herbicides), grading and packaging.

Model Agricultural Produce Marketing Act

Structural and institutional reforms are necessary to make the present agricultural marketing system competitive and efficient. The Model Agricultural Produce Marketing Act is an attempt to overcome some of the major bottlenecks in the present system and to promote competitive marketing from farm to the consumer's plate.

The Act highlights the need for an alternative marketing system to promote direct marketing, smooth raw material supplies to agro-processing industries, competitive trading, organized retailing, information exchange and adoption of innovative marketing systems and technologies (GOI, 2003).

The key recommendations include:

- Provision of 'Special Markets' or 'Special Commodities Markets' in any market area for specified agricultural commodities to be operated in addition to existing markets.
- No compulsion on the growers to sell their produce through existing markets administered by the Agricultural Produce Market Committee (APMC).
- Formation of Farmers Association for bulk marketing
- Pledge financing
- System of negotiable warehouse receipts
- Forward and futures contracts
- Promotion of information technology in agricultural marketing
- Extension and training services need to focus on assisting small and marginal farmers in (a) marketing of their produce, (b) advice on production planning, (c) market information, (d) alternate marketing channels, (e) improved marketing practices including grading and packing, and (f) advantages of group marketing.

Once the Model Marketing Act is effectively implemented, it will enable compulsory registration of all contract farming sponsors, recording of contract farming agreements and resolution of disputes arising out of such agreements. The Act has provision to exempt from levy of market fee on produce covered by contract farming agreements. Under the Act, new markets are expected to be established by the private sector and other legal entities, and farmers are free to sell their produce in any market. It will create competition between the existing and new markets, thus ensuring a fair deal to the producers. Thus, the Act is expected to promote public-private partnership in the management of agricultural markets.

Conclusions and Policy Implications

Unfolding of market liberalization and globalization is causing significant changes in agriculture and agri-food markets. Agriculture is diversifying towards high-value food commodities, and food-marketing system is moving towards vertical coordination. In order to ensure that the benefits of emerging opportunities in agriculture are not confined to large producers but are also reaped by smallholders who are hard hit by the ongoing agrarian crisis, there is a greater need for strong linkages between production and markets and an enabling policy environment for strengthening the supply chain and value addition.

Integration of smallholders on the supply chain is a major challenge. Exclusion of smallholders from the supply chain is politically unacceptable and socially undesirable. Their integration requires collectivization into cooperatives and self-help groups or intermediary contracts, which reduce transaction costs to both firm and farmers. The government should facilitate such institutions.

Production of most of the high-value agricultural commodities is capital-and information/knowledge-intensive and riskier, while smallholders lack access to capital, improved technologies, quality inputs, and extension services needed for entry into high-value segment. It is, therefore, essential to strengthen institutional mechanisms that improve smallholders' access to credit, insurance, technology and support services.

Expanding agribusiness will confront infrastructural constraints such as poor roads, transportation and storage facilities, and erratic electricity supply. Empirical evidence has shown a lower concentration of high-value agriculture in areas with poor road network (Parthasarathy Rao *et al.*, 2006). Investment is required in public infrastructure not only to facilitate market linkages but also to induce private investment in cold chains and food processing.

The model marketing act, though laudable, needs to be integrated into the existing marketing act by the state governments. Although 7-8 state governments have come forward and amended the marketing act by including the provisions of the model act, implementation of these provisions at the ground level is slow. Hence, steps need to be taken for the speedier implementation that would ultimately benefit the smallholders under the changing agricultural scenario.

References

Aksoy, M.A. (2005) The evolution of agricultural trade flows. In: *Global Agricultural Trade and Developing Countries*, Eds: M.A. Aksoy and J.C. Beghin. The World Bank, Washington D.C.

Bakshi, K., D. Roy and A. Thorat (2006) Small they may be and Indian farmers they are but export they can—The case of Mahagrapes farmers in India. Paper presented at the Workshop *From Plate to Plough—Agricultural Diversification and its Implications for Smallholder*, organized by the International Food Policy Research Institute, New Delhi office, and Institute of Economic Growth, New Delhi, India on 20-21 September.

Birthal, P.S., P.K. Joshi and Ashok Gulati (2005) *Vertical Coordination in High-value Food Commodities: Implications for Smallholders*. MTID Discussion Paper 85. International Food Policy Research Institute, Washington D.C.

Birthal, P.S., A.K. Jha, M. Tiongco, C. Delgado, C. Narrod and P.K. Joshi (2006) Equitable intensification of smallholder market-oriented dairy production in India through contract farming. Annex I of the IFPRI-FAO final report *Contract Farming of Milk and Poultry in India: Partnerships to Promote the Environment-friendly and Equitable Intensification of Smallholder Livestock Production*. International food Policy Research Institute, Washington DC, USA.

Birthal, P.S. and P.K. Joshi (2007) Institutional innovations for improving smallholders participation in high-value agriculture: A case of fruit and vegetable growers' associations. *Quarterly Journal of International Agriculture*, 46(1): 49-68.

CFC-FAO-ICRISAT (2006) Comprehensive report on innovations in supply chain and suitable supply and marketing systems for sorghum and bajra. Unpublished report.

Diaz-Bonilla, E. and L. Rccca (2000) Tradc and agro-industrialization in developing countries. *Agricultural Economics*, 23(3): 219-230.

Eaton, C. and A.W. Shepherd (2001) Contract farming: Partnerships for growth. *FA O Agricultural Services Bulletin 145*. Food and Agriculture Organization, Rome, Italy.

Erappa, S. (2006) *Contract Farming in Karnataka: A Boon or a Bane?* Research Report 9/ADRT 113. Agricultural Development and Rural Transformation Centre, Institute for Social and Economic Change, Bangalore.

FAOSTAT (2007) http://faostat.fao.org. Food and Agriculture Organization, Rome, Italy.

Gandhi, V. and N.V. Namboodiri (2002) *Fruit and Vegetable Marketing and its Efficiency in India: A Study in Wholesale Markets in Ahmedabad Area*. Memio. Indian Institute of Management, Ahmedabad.

GOI (Government of India) (1999) *Cultivation Practices in India*. Report No. 451 (54/31/3). National Sample Survey Organization, Ministry of Statistics and Programme Implementation, New Delhi.

GOI (Government of India) (2003) *Salient Features of the Model Act on Agricultural Marketing*. Directorate of Marketing and Inspection, Ministry of Agriculture, Government of India. Available at http://agmarknet.nic.in/reforms.htm. Accessed on 23 May 2005.

GOI (Government of India) (2005) *Income, Expenditure and Productive Assets of Farm Households*. Report No. 497 (59/33/5). National Sample Survey Organization, Ministry of Statistics and Programme Implementation, Government of India, New Delhi.

GOI (Government of India) (2006a) *Level and Pattern of Consumer Expenditure, 2004-05*. Report No. 508 (61/1.0/1). National Sample Survey Organization, Central Statistical Organization, Ministry of Statistics and Programme Implementation, New Delhi.

GOI (Government of India) (2006b) *Some Aspects of Landholding in India, 2002–03*. Report No. 492 (59/18.1/3). National Sample Survey Organization, Central Statistical Organization, Ministry of Statistics and Programme Implementation, New Delhi.

Joshi, P.K., Ashok Gulati, Laxmi Tewari and P.S. Birthal (2003) *Agricultural Diversification in South Asia: Patterns, Determinants and Policy Implications*. Markets, Trade and Institutions Division (MTID) Discussion Paper No. 57, IFPRI, Washington DC.

Joshi, P.K., Ashok Gulati, P.S. Birthal and Laxmi Tewari (2004) Diversification of Agriculture in South Asia. *Economic and Political Weekly*, 12-19 June.

Kleith, U., S. Bala Ravi, B. Dayakar Rao and Yoganand (2002) *Industrial Utilization of Sorghum in India*. Working Paper Series No.4., International Crops Research Institute for the Semi Arid Tropics, Patancheru 502 324, Andhra Pradesh 44 pp.

Kumar, P. (2006) Contract farming through agribusiness and state corporation. *Economic and Political Weekly*, 41(52): 30 December.

Marshland, N. and P. Parthasarathy Rao (1999) *Marketing of Rainy and Postrainy Season Sorghum in Andhra Pradesh, Karnataka, and Maharashtra*, Working Paper Series No.1, International Crops Research Institute for the Semi-Arid Tropics, Patancheru, Andhra Pradesh.

NDDB (2005) *Annual Report*. National Dairy Development Board, Anand, Gujarat.

Parthasarathy Rao, P., P.S. Birthal and P.K. Joshi (2006) Diversification towards high-value Agriculture: Role of urbanization and infrastructure. *Economic and Political Weekly*, 30 June:.2747-2753.

Ramaswami, B., P.S. Birthal and P.K. Joshi (2005) *Efficiency and Distribution in Contract Farming: The Case of Indian Poultry Growers*. MTID Discussion Paper No. 91. International Food Policy Research Institute, Washington DC, USA.

Ravi, C. and D. Roy (2006) Consumption patterns and food demand projection: A regional analysis. Paper presented at the Workshop *From Plate to Plough—Agricultural Diversification and its Implications for Smallholders* organized by the International Food Policy Research Institute, New Delhi Office, and Institute of Economic Growth, New Delhi, India, 20-21 September.

9

Rise of Food Retail Chains in India: Prospect and Impact

P.G. Chengappa and Vijayalakshmi Dega

'Food Retail Chains' or 'Supermarkets', which were traditionally viewed by the development economists, policymakers, and practitioners as niche markets for rich consumers in the capital cities of the countries have transformed to places of destination shopping in several countries. Rapid urbanization, increase in the number of households with working women, and rapid increase in per capita income have been identified as being responsible for the explosive growth of supermarkets. These demand-side factors together with shifts in policies relating to opening up of economies and reduction in restrictions on Foreign Direct Investment (FDI) were responsible for transforming the retail sector in several developing countries (Reardon *et al.*, 2002). These changes, marked during the past 5-10 years, have witnessed the transformation of agri-food markets at different rates and depths across regions and countries. Supermarkets have brought changes in the agri-food systems in several countries (Reardon and Berdegue, 2002; Weatherspoon and Reardon, 2003; Reardon *et al.*, 2003; 2004). The entry of supermarkets has introduced several changes in the production and marketing of agricultural crops, especially fruits and vegetables. To achieve their objectives of trading in quality and safe products and reducing costs and increase volumes, supermarkets are in the process of shifting away from the old procurement models, i.e. procuring from the traditional wholesalers, towards a new procurement systems based on enlisting specialized procurement agents, centralized procurement, and preferred suppliers, all of which help them to trade only in high quality and safe products (Reardon *et al.*, 2004).

The changes introduced by organized food retailing have significant opportunities for agricultural producers (Schwentesius and Gómez, 2002; Weatherspoon and Reardon, 2003; Reardon *et al.*, 2003; 2004; 2005). Opportunities for realizing higher incomes, employment generation and risk minimization have been indicated in most instances. These linkages, however, could also mean larger investments on the part of farmers to meet the tougher quality standards imposed by the food retail chains. Lack of capital could lead to exclusion of small farmers from an efficient marketing system (Reardon and Berdegué, 2002; Weatherspoon and Reardon, 2003). Farmers also face the threat of "arm twisting" tactics adopted by the food retail chains.

This Chapter has examined the facets of food retail chain, including the pattern in the rise of supermarkets in India, factors influencing or hampering their rapid diffusion, procurement systems of supermarkets and impact of procurement system on producers' resource-use pattern and income. A primary survey, during July-December 2004, of four major food retail chains, namely, *Nilgiris, Food World* (now Spencers), *FabMall, Subhiksha,* and information gathered from various categories of respondents, such as consumers and traditional retailers, forms the basis of this study.

Pattern of Development of Food Retail Chains in India

There has been rapid expansion of organized food retail chains in India during the past five years. The post-liberalization era in the 1990s provided an environment conducive for the establishment of food retail chains. During this period, Indian food retailing sector witnessed the emergence of international formats of retailing, such as the supermarkets/grocery chains, convenience stores and fast-food chains. In 1996, there were only a handful of supermarkets in major cities of India. These stores had an area ranging from 3000 sq ft to 5000 sq ft, and offered a wide range of groceries, snacks, processed foods, personal care products, cosmetics, etc. It has been reported that there were about 385 (in 2003) such supermarkets operating all over the country, mostly with one outlet, while there were 14 companies that ran departmental stores, and two hypermarkets. However, recent years have witnessed a number of corporate players, such as the Spencers', Piramals, the Tatas, the Rahejas, ITC, Pantaloon, and the RPG group entering the organized food retail sector in India (CII, 2003). Despite these developments, there are no national players mostly due to capital and logistic constraints.

The modern food retail chains are self-service stores, with wider range of products, have their own brand (in-house brand) for groceries and home products, with smaller sections devoted to fresh fruits and vegetables, bakery products, dairy products as well as frozen meat. A typical food retail chain has about 6000 Stock Keeping Units (SKU). The food retail chains evolved from being low retail technology to having IT-based technology for inventory management, billing, accounts, customer care, etc. Gross margins of these chains ranged from 14 per cent to 16 per cent, which are higher than from the other formats due to ability to source goods from manufacturers at lower prices for bulk purchases as well as selling, advertising space and special in-store promotion by manufacturers. These stores cater to the population with higher incomes who seek a wider choice of products, possessing storage space and means of transport facilities for larger purchases. Organized food retail chains being of recent origin account for less than 2 per cent of the food retailing industry in India (Table 9.1), but with a projected growth rate of 8.3 per cent per annum, it is expected to appropriate a share of 20 per cent of total food retailing by 2008 (KnightFrank, 2003).

Table 9.1: Share of Organized Food Retailing in India

Food retailing	*1999*	*2002*	*2005**
Total food retailing (USD billion)	159	180	225
Organized food retailing (USD billion)	1.1	3.2	7.0
Share of organized food retailing, %	0.70	1.89	3.20

*projected

Source: Gupta and Bhalotia (2003), www.indiainfoline.com

Formats of Organized Food Retail Chains

Food retailing formats in India differ from the existing global formats. The size of the outlets is much smaller compared to those in other countries. The reasons for this being high cost of real estate, low investment capacity among domestic players, restrictions on FDIs, lack of technical know-how and competition from traditional retailers. Retail formats in the food retail sector in India can be categorized as:

Super Centre–Huge area, sells an assortment of food and non-food items; has supermarkets, departmental stores, music, food and entertainment areas (e.g., Family Mart 100,000 sq ft)

Hyper Markets–Stores that take the benefits of the superstore even further, using their greater size to give the customer a wider range and depth of products (e.g., Big Bazaar, Giant)

Large Supermarkets (typically 3,500-5,000 sq.ft)–Large, self-service stores that carry a complete line of food products as well as other convenience items (e.g., Food World, Nilgiris)

Discount/Shopping List Grocers–Sell their products below MRP (e.g. 'Subhiksha' sells at 1-9 per cent below MRP)

Pattern of Procurement by Food Retail Chains

Food retail chains continue to depend on traditional food grain marketing channels for the supply of food grains, and have not established backward linkages for their procurement directly from farmers. In the case of manufactured commodities, such as detergents, processed foods, beverages, etc., the retail chains procure them directly from manufacturers, reducing the involvement of middlemen. Traditional wholesalers continue to play an important role in the procurement system of fresh fruits and vegetables (FFV) of all major food retail chains. However, one of the food retail chains, Spencer's (previously Food World) has established an innovative mechanism of backward linkages with farmers for direct procurement of fresh fruits and vegetables. Food retail chains indicated that they procured their requirements of food grains (cereals and pulses) from the regulated market yards (APMC yards) and fruits and vegetables from the regulated wholesale market and newly established SAFAL market. The retail chains in the food sector are forced to buy at the regulated markets due to the existence of APMC law. They are neither able to compress their supply chain (Figure 9.1) nor make any significant improvement in the marketing infrastructure due to the restrictions. The existing laws governing the marketing of agricultural produce in the country preempt efforts to shorten the supply chain even though the stated objectives of the policy of the government were ostensible to accomplish such a reduction.

Fruits and vegetable marketing in India is characterized by the domination of wholesalers, commission agents and pre-harvest contractors, while retail trade is decentralized with the presence of

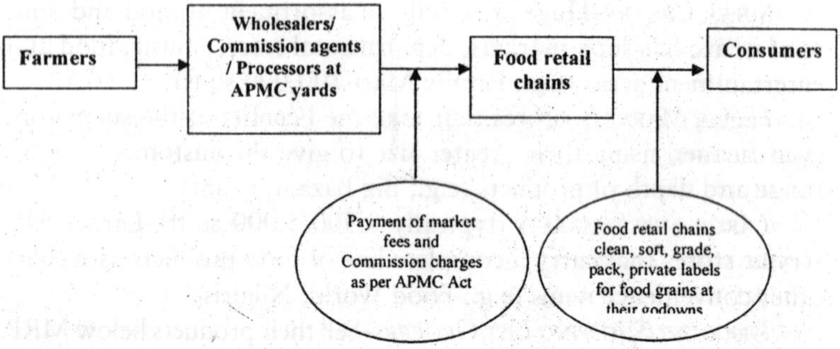

Figure 9.1: Supply Chain for Foodgrains of Food Retail Chains

millions of small retail shops, pushcart vendors, and roadside hawkers. Food retail chains being new entrants in the food-retailing scenario have not been able to penetrate the FFV marketing scenario, leading to small FFV business volumes, thus precluding the need to have direct linkages with farmers for procurement of FFV. Only one retail chain, (Spencer's, formerly Food World), has initiated a new institutional arrangement for procuring FFV directly from farmers through their consolidation centre at Hoskote (Bangalore rural district, Karnataka).

Linking Farmers to Retail Food Chain: Case of Spencer's

Spencer's, in 1996 established (effectively functioning from 2002) a consolidation centre in Hoskote (in vegetable producting belt) near Bangalore city. Its area of operation extends to a radius of 50-80 km. This Centre procures FFV directly from farmers, which include about 163 locally grown varieties of vegetables (some exotic ones), and to a small extent, fruits. At present, the number of farmers registered (as vendors) at the consolidation centre is small (19 regular suppliers and 11 seasonal suppliers). The popularity of this centre can be measured through the number of farmers in the waiting list (more than 600).

The Bangalore centre handles about 20 tonnes a day (600 Mt a month). It plans to reach a procurement target of 5000 tonnes per month by 2007. At present, the consolidation centre meets only about 50 per cent of Spencer's requirement of fruits and vegetables, while the remaining is procured locally from the SAFAL market established by the National Dairy Development Board (NDDB) through a consolidator.

Prior to introduction of this system, Spencer's depended on wholesalers in the traditional marketing channel, with no control over the quality and price of the produce. In this model, farmers registered with Spencer's were known as 'vendors', under whom a group (about 10) of farmers (independently) cultivated and supplied fruits and vegetables (Figure 9.2). The selection of vendors is based on their business management skills as supply to the centre involves more formal transaction methods as well as stringent delivery conditions, frequency of supply, quality standards for the product, etc. So far, relationships have been informal, with no written contracts, and are based on oral confirmations of volumes to be delivered. The registered farmer-vendors collect the produce of other farmer-members and deliver it to the consolidation centre; quality controls in production and packaging being the responsibility of the farmer-vendor themselves. The concept adopted by Spencer's is *"Farmer to the Market"*, in which produce is graded and packed in the form required by consumers ('Ready to Retail Products'). For example, carrots and beans are available in 250 g packs, and green chillies in the packs of 100 g, etc. By shifting the responsibilities of cleaning, sorting, grading and packing to the farmer-vendors, the food chain is able to reduce the transaction costs. At the centre, packed produce are bar coded and transported to the central warehouse in Bangalore, from where these are further transported to other south India cities, viz. Chennai, Hyderabad, and Thiruvanthapuram (located more than 300–500 km away from this consolidation centre).

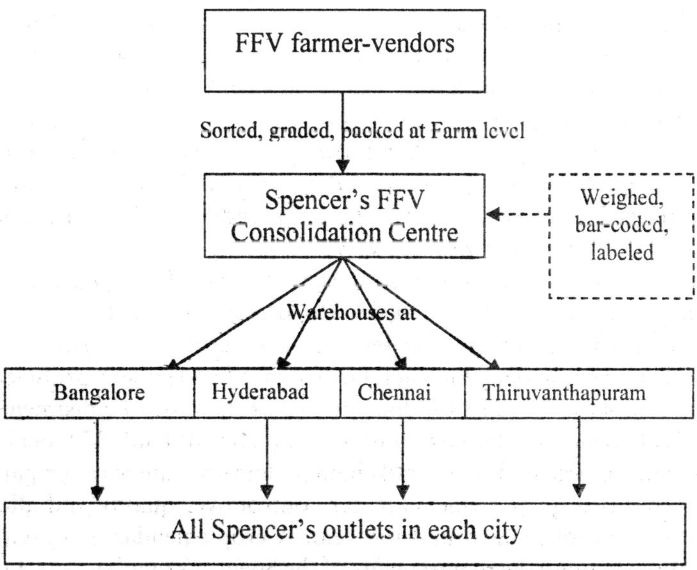

Figure 9.2: Spencer's Fresh Fruits and Vegetables (FFV) Supply Chain

This new institutional arrangement by Spencer's has helped the farmers to break from the stranglehold of traditional brokers/ wholesalers/commission agents. Direct sourcing from farmers has provided the retail chain with increased control over quality, supply reliability and price stability. Unlike in other countries such as Kenya, where supermarkets work to deal with fewer and larger suppliers (Neven and Reardon, 2004), in India, Spencer's have emphasized on having supplies from smaller farmers with small and medium size of landholding (Table 9.3).

Table 9.3: Distribution of Sample Farmers Across Landholding Sizes

(in per cent)

Category of farmers	*Spencer's*	*Non-Spencer's*	*Total*
Small	3 (15.79)	28 (93.33)	31 (63.27)
Medium	9 (47.37)	2 (6.67)	11 (22.44)
Large	7 (36.84)	0 (0.00)	7 (14.29)
Total	19 (100.00)	30 (100.00)	49 (100.00)

Source: Field survey conducted by University of Agricultural Sciences, Bangalore in 2004-05

Dependence on smaller farmers has certain intrinsic advantages such as: (i) smaller farms have comparative advantages in production of certain crops, such as mint, coriander, exotic vegetables, etc., (ii) availability of family labour which is low cost and self-supervisory as opposed to paid labour on large farms, which needs to be supervised, (iii) management of diseases and hence reduction in large crop losses, and (iv) smaller farms provide effective response to problems of land scarcity. To meet the unexpected increase in demand, Spencer's have a list of stand-by farmers (under each vendor-farmer), from whom they can procure FFV.

Changes in Cultivation Practices Introduced by Spencer's: The consolidation centre of Spencer's has brought about changes in the way crops, particularly vegetables are cultivated. Crops to be cultivated are assigned to each farmer based on farmer's proficiency and history of production documented at the time of enlistment. A typical farmer-vendor has around 7.45 (acres) of land, 87.9 per cent of which is irrigated. The retail chain prefers investments in irrigation systems as it gives farmers greater control over quality and allows them to produce all round the year. A crop calendar is drawn up keeping in mind the requirements of the Spencer's retail outlets. Once the crop to be cultivated has been decided, farmers are provided with a package of Good Agricultural Practices (GAP). This package ensures the optimum use of resources with emphasis on minimum use of pesticides. The vendor-leader ensures that the practices are strictly adhered. Direct supplies by farms also allow the centre to inspect farm and growing practices on a first-hand basis. There is no formal contract between the firm and the producer under this arrangement. The centre does not supply any production inputs nor do they formally agree to procure the produce, which makes the farmers risk-bearers. The marketing arrangement by Spencer's, however, reduces the market risks (for good quality produce) and transaction costs to the farmers. Spencer's also have no system of providing production credit to the farmers. Spencer's help the farmers in procuring inputs from suppliers at reduced rates. Technical guidance on aspects like the time of planting, crop production and management, harvest time, quantity to be harvested per acre, etc., to ensure quality and marketability, are provided by the consolidation centre.

Generally, the centre procures the entire quantity of FFV supplied by the vendors, except in cases where the specified quality requirement is not met. During the initial stages of establishment, the percentage of rejection in procurement from farmers was high because the farmers were not accustomed to producing good quality produce in a scientific manner. The large-scale rejection of their produce failing to meet the quality specifications led the farmers to change their cultivation practices, following which the rate of rejection reduced and now stands at eight per cent. The impact of the Good Agricultural Practices (GAP) and production practices such as, staggered sowing introduced by the centre has led to increase the intensity of cultivation as well as increase the production.

Pricing Policy: Spencer's follow a mix of policies for determining the prices to be paid to the farmers. Benchmarks include prices prevailing at SAFAL market, HOPCOMS and the major wholesale market (Krishna Rajendra Market) in Bangalore. For exotic vegetables, they follow an annual contract (price) system. It was found that the farmers preferred supplying their produce to the consolidation centre, as it provided them stable prices and assured market, compared to the highly volatile prices at the wholesale market. Farmer-vendors who supply low volume-high margin products are paid in cash once in 2 to 3 weeks, while those supplying high volume-low margin products are paid once in 7-15 day period.

This delay in payment was because of the existing tedious accounting practices being followed by the retail chain.

Farmer-vendors associated with the food retail chain cultivate in addition to the traditional vegetables grown in the region, other crops (Table 9.4), including many exotic ones such as broccoli, iceberg lettuce, parsley, leek, red cabbage, chinese cabbage, colour capsicum, green onion, turnip, basil and table radish. This is due to the fact that marketing risks are taken care-of by the new institutional arrangement.

Table 9.4: Crops Cultivated by the Two Groups of Farmers

Spencer's farmer-vendors	Non-Spencer's farmers
Ash gourd, Baby corn, Beet root, Bottle gourd, Brinjal (egg plant), Capsicum, Cucumber, Musk melon, Pumpkin, Ridge gourd, Bottle gourd, Cabbage, Carrot, Cauliflower, Double beans, Green Chillies, Little gourd, Potato, Broccoli, Iceberg lettuce, Parsley, Leek, Red cabbage, Chinese cabbage, Colour capsicum, Green onion, Turnip, Basil and Table radish.	Bottle gourd, Cabbage, Carrot, Cauliflower, Double beans, Green Chillies, Little gourd, Potato

Source: Field survey conducted by University of Agricultural Sciences, Bangalore in 2004-05

A Comparison of Profits and Transactions Costs: The differences in profits and transaction costs used as indicators of the performance of the institutional arrangement in the marketing of agricultural commodities have revealed noticeable differences in the net returns per tonne (Table 9.5). The difference was substantial in the case of cauliflower. The increase in net returns was the highest for cauliflower (57 per cent), followed by tomato (43 per cent). The figures for cabbage and carrot were 47 per cent and 13 per cent, respectively.

The farmers enjoyed lower costs of production by following the recommended Good Agricultural Practices. Reductions were seen in the use of quantities of seeds, chemical fertilizers and plant protection chemicals. Another major advantage for farmers with a tie-up with Spencer's was the reduction in transaction costs, due to the absence of intermediaries in the supply chain The reductions in transaction costs for Spencer's farmers were enormous, with figures ranging from 48.4 per cent for tomato to 85 per cent for cauliflower. Table 9.6 which gives various components of transactions incurred by both groups of farmers, shows that reduction in transactions costs under the new institutional arrangement (Spencer's) was due to the absence of

Table 9.5: A Comparison of Unit Cost of Production and Profit in Production of Vegetable Crops by Spencers and Non-Spencers Farmers

Crops	Cost of production (Rs/tonne)		Transaction costs (Rs/tonne)		Total Cost (Rs/tonne)		Net return (Rs/tonne)		Increase (%)
	Spencers	Non-Spencers	Spencers	Non-Spencers	Spencers	Non-Spencers	Spencers	Non-Spencers	
Cabbage	1267	1427	206 (14.01)	0	1473	1427	2016	1072	46.83
Cauliflower	3338	3457	214 (6.02)	1430 (29.26)	3552	4887	4878	2112	56.70
Carrot	3223	2861	801 (19.91)	0	4025	2861	10475	9138	12.76
Tomato	1643	1809	652 (28.41)	1265 (41.14)	2295	3074	4244	2425	42.85

Notes: Figures within parentheses indicate percentages of transaction cost to total cost

Source: UAS Field Survey (2004)

Table 9.6: Components of Transaction Costs for Vegetables Marketed Through Traditional Mode and Spencer's Supply Chain

Details	*Cabbage*		*Cauliflower*		*Carrot*		*Tomato*	
	Spencer's	*Traditional*	*Spencer's*	*Traditional*	*Spencer's*	*Traditional*	*Spencer's*	*Traditional*
1. Transaction cost of farmer								
a. Labour	104	-	124	50	225	-	140	80
b. Packing material	-	-	-	142	400	-	400	125
c. Transportation	76	-	64	400	150	-	100	400
d. Commission	-	-	-	700	-	-	-	500
e. Personal expenses	8	-	7	12	8	-	3.504	20
f. Loading and Unloading	-	-	-	100	-	-	-	100
g. Non pecuniary costs	19		18	25	18	-	8	40
Sub-total	207	-	213	1429	801	-	652	1265
Producers sale price/ Wholesalers/Retailers purchase price	3490	2500	8430	7000	14500	12000	6540	5500
Net price received by the Farmer	**3284**	**2500**	**8216**	**5570**	**13699**	**12000**	**5988**	**4235**
2. Transaction cost of wholesaler/FW consolidation centre								
a. Packing cost	-	50	-	50	-	62	-	75
b. Hamali charges	-	50	-	50	-	62	-	75
c. Personal expenses	-	50	-	50	-	55	-	50
d. Transportation	280	-	280	-	280	-	280	-
Sub-total	280	150	280	150	280	180	280	200
Wholesalers margin	-	1200	-	850	-	3440	-	1300
Wholesalers sale price	-	**4000**	-	**8000**	-	**15800**	-	**7000**

3. Transaction cost of the retailer/Food world warehouse								
a. Shop rent	–	100	–	100	–	120	–	100
b. Wastage	–	–	–	–	–	–	–	–
c. Personal expenses	–	450	–	500	–	400	–	450
d. Transportation	425		425	-	425	–	425	–
Sub-total	–	550	425	600	425	520	425	550
Retailers margin	–	1450	–	1400	–	3200	–	1450
Retailers sale price	**6500**	**6000**	**10900**	**10000**	**21500**	**19520**	**11000**	**9000**
Producer's share in consumer rupee (%)	**50.51**	**38.33**	**75.38**	**55.70**	**63.72**	**61.46**	**53.53**	**47.06**

Source: UAS Field survey, 2004

unloading charges, commissions paid to the wholesalers, reductions in transportation costs, labour charges and other costs. The reductions in transportation costs were due to the proximity of the consolidation centre to the farmer-vendor's fields. The farmer-vendors associated with Spencer's however, had to incur additional labour charges needed for cleaning, washing, grading and packaging of the vegetables as per the specification of the retail chain. Labour employment was also more as the crop harvest had to be staggered to meet the indent specification. These costs were, however, offset by the higher prices offered by the retail chain.

Prices prevailing during the study period, presented in Table 9.7, reveal that non-Spencer-farmers selling in the regulated markets (wholesalers/commission agents) and through pre-harvest contractors received 4-40 per cent lesser prices as compared to prices received at Spencer's consolidation centre. The pre-harvest contractors particularly paid much lesser prices to farmers. However, farmers did not incur transaction costs on harvesting and transporting produce to the market when sold through pre-harvest contractors.

Table 9.7: Prices Offered Under Different Marketing Arrangements

Crop	*Prices offered (Rs/t) Spencer's Consolidation Centre*	*Wholesalers/CAs at regulated market*	*Pre-harvest contractor*
Cabbage	3490	3000 (-16.33)	2500 (-39.60)
Cauliflower	8430	7000 (-20.43)	-
Carrot	14500	14000 (-3.57)	12000 (-20.83)
Tomato	6540	5500 (-18.91)	-

Note: Figures within the parentheses indicate percentage of reduction in prices paid by various agents as compared Spencer's FV

Source: UAS Field Survey (2004)

Factors Influencing Participation of Farmers in the New Institutional Arrangement: After about 10 years of operation, only 19 farmers–vendors (in all about 200 farmers) have been enlisted by the FVC, which indicates that certain criteria have been adopted for accepting farmers as vendors in this system. To identify factors that influenced the participation of farmers in the new institutional arrangement for procurement of FFV by the food retail chain, a logit regression model was fitted to the data to predict the factors that induce registration with the retail chain. The structural form of the logit model is given by Equation (1):

$$L_i = \alpha_i + \beta_i X_i + \mu_i \ldots \qquad (1)$$

where, L_i is the dependent variable taking the value 1 if the respondent is a registered vendor with Spencer's and 0, otherwise; X_is is a vector of independent variables such as age, education, size of landholding, family size, number of bore-wells owned; α_i and β_i are the estimated parameters; and μ_i is the error-term. It is hypothesized that as education, family size, size of landholding and ownership of a larger number of bore-wells increase, the probability of their participation in the new system of procurement increases, while age is hypothesized to have a negative influence on participation.

The results (Table 9.8) indicate that ownership of bore well was the most important factor that improved the odds of a farmer participating in the new system, which was as per expectations. Food retail chains preferred farmers with assured irrigation, as this gave farmers better control over production activities and helped in the production of quality produce at the same time assured supplies of FFV to the food chain. Education of farmers and size of landholding were other important factors inducing adoption. The prediction accuracy of the model revealed that 95 per cent of the observations were correctly predicted by the results of the logit regression.

Table 9.8: Factors Determining Farmer's Participation in the Food Retail Chain

Factor	*Hypothesized relationship with Y*	*Regression co-efficient*	*Odds ratio*
Age, years	-	-0.0814 (0.06)	0.92
Bore-well, No.	+	4.2414 (44.14)	69.51
Education, years	+	0.5944* (0.27)	1.81
Family size, No.	+/-	-0.1407 (0.2593)	0.87
Size of landholding, acres	+	0.9902* (0.36)	2.69
Constant		-7.3298 (44.23)	
Chi-square value		45.04**	
No. of observations		50	

Note: Figures within parentheses indicate standard errors, * significant at 5 per cent probability level, ** significant at 1 per cent probability level. log Likelihood= 20.389, Cox and Snell R^2 = 0.601, Nagelkerke R^2 =0.816

Source: UAS Field Survey (2004)

Conclusions

Studies world over (Reardon and Berdegue, 2002; Weatherspoon and Reardon, 2003; Reardon *et al.*, 2003; 2004; Balsevich *et al.* 2003; Hu *et al.*, 2004; Sheperd, 2005) have indicated that the emergence of food retail chains has introduced several changes such as centralization and rationalization, move towards specialized wholesalers and logistic firms, and gradual movement towards preferred suppliers as well as the rise of private standards. In India, since organized food retail is still in its infancy, these transformations have been slow. This is also due to the impediments arising from the archaic legislations, especially those related to the procurement of primary agricultural commodities (APMC). Nevertheless, changes such as centralization in procurement by retail chains to increase efficiency and reduce transaction costs are happening. Food retail chains have also moved towards preferred suppliers known as vendors and consolidators who are responsible for the procurement and supply of specified quality of products. Recently, there has been move towards rise of private standards. Most of the food retail chains in India have set their own standards for the quality of food grains and other grocery items (manufactured products meet the standards set in the country). The retail chains also have introduced private labels for these items, many of which are priced slightly lower than the non-private label products. So far, however, in India, there has been no move towards rationalization of procurement, mainly due to the separate agricultural legislation for each state in the country. There has also been no move towards the outsourcing of wholesale distribution and other logistic functions to specialized firms.

Regarding linkages with farmers, it has been found that currently, all food retail chains, except one—Spencer's procure their requirements from wholesale markets, implying that most of the food retail chains have no direct backward linkages with farmers. Spencer's is the only retail chain that operates a fruits and vegetables (FFV) consolidation centre. The current low sale volumes of FFV stocked by food retail chains do not warrant tie-ups with farmers. The food retail chains however, have indicated that they are open to the idea of linkages with farmers.

Note

1. The difference in net returns for cabbage would be higher (64 per cent) if the farmers in the traditional supply chain had harvested and transported the crops to the market themselves. But, due to reasons such as high demand for cabbage and high cost of harvesting and transportation, in 90 per cent of the cases, pre-harvest contractors visited the farmers' fields and fixed a price for the crop and they themselves harvested the crop, due to which farmers did not have any problem in disposing the produce.

References

Balsevich, F., J.A. Berdegue, L. Flores, D. Mainville and T. Reardon (2003), Supermarkets and produce quality and safety standards in Latin America. *American Journal of Agricultural Economics*, 85(5): 1147-54.

CII (2003), *Retail Scenario in India*, Report, Confederation of Indian Industries, New Delhi.

Gupta, S. and N. Bhalotia (2003), *Retailing in India-Past, Present and Future.* Indian Institute of Foreign Trade, (www.indiainfoline.com).

Hu, D., R. Reardon, S. Rozelle, P. Timmer and H. Want (2004), The emergence of supermarkets with Chinese charactersitics: Challenges and opportunities for China's agricultural development. *Development Policy Review*, 22(5): 557-586.

KnightFrank (2003), *India–Retail Review.* http://www.knightfrank global.com.

Naven, D. and T. Reardon (2004), *Kenyan Supermarkets and Horticultural Farm Sector Development*, Contributed Paper Prepared for Presentation at the International Association of Agricultural Economists Conference, Gold Coast, Australia.

Reardon, T. and J.A. Berdegué (2002), The rapid rise of supermarkets in Latin America—Challenges and opportunities for development. *Development Policy Review*, 20(4):371-388.

Reardon, T., J.A. Berdegué and C.P. Timmer (2005), Supermarketization of the "Emerging Markets" of the Pacific Rim: Development and trade implications, *Journal of Food Distribution Research*, 36 (1).

Reardon, T., J.A. Berdegué and J. Farrington (2002), *Supermarkets and Farming in Latin America, Pointing Direction for Elsewhere?* Natural Resource Perspectives, No. 81, ODI (www.odi.org/uk/nrp/81.pdf)

Reardon, T.A., J.A. Berdegue, M. Lundy, P. Schutz, F. Balsevich, R. Hernandez, E. Perez, P. Jano and H. Wang (2004), *Supermarkets and Rural Livelihoods: A Research Method.* Staff paper, Michigan State University, September.

Reardon, T., P.C. Peter Timmer and J.A. Berdegué (2003), The rise of supermarkets in Latin America and Asia: Implications for international markets for fruits and vegetables. Chapter In: *Global Markets for High Value Food Products*, Eds: Anita Regmi and Mark Gehlhar, Agriculture Information Bulletin, USDA-ERS.

Schwentesius, R. and M.A. Gómez (2002), Supermarkets in Mexico: Impact on horticulture systems, *Development Policy Review*, 20(2): 487-502.

Shepherd, A. (2005), *The Implications of Supermarket Development for Horticultural Farmers and Traditional Marketing Systems in Asia*. Food and Agriculture Organization, Rome.

Weatherspoon, D.D. and T. Reardon (2003), The rise of supermarkets in Africa: Implications for agrifood systems and rural poor. *Development Policy Review*, 21 (3): 333-355.

10

Rural Non-Farm Sector as an Option to Alleviate Farmers Distress

Brajesh Jha

Although agriculture remains the mainstay of more than 72 per cent of rural workforce, several studies in recent years have shown the status of Indian farmers as poor. There may be innumerable reasons for farmers' distress, some of which are periodic in nature[1]. An unviable size of landholdings is at the core of such distress-related issues in the rural sector. The majority of Indian farmers now require off-farm income for sustained rural livelihood. This is possible with a robust growth of rural non-farm sector (RNFS). Economic opportunities in non-farm sector have increased, but the rate of growth of employment is much lower than the growth of labour-force in the rural India. A *lopsided* growth of the non-farm sector is causing problems, like rural-urban migration. An enquiry into RNFS has revealed that sustained growth and performance of the sector depend on a variety of factors as it comprises diverse activities. These factors induce development-or distress-led rural diversification depending on the kind of impetus it provides to rural economy.

Several researchers have found that growth in agriculture, urbanization, infrastructure and human resource related parameters stimulate development-induced rural diversification in the country. Jha (2007, 2006a) has also found that the importance of agriculture in rural diversification has decreased in recent decade; though agriculture—induced rural diversification has better spatial spread of welfare in a region. The manufacturing is a leading component of the non-farm sector. Employment growth in manufacturing has increased in the recent decade (Rani and Unni, 2004), but there are also evidences of manufacturing shifting away from the rural sector (Jha 2006a). The recent NSS quinquennial survey results have suggested that construction, trade, transport and business services are the new source of RNFS growth in the country. The present Chapter has examined the extent to which employment growth in these sectors can drive rural diversification in India and ascertain how growth in rural livelihood can be sustained.

Pattern and Potential of Rural Diversification

Rural diversification, referred as an increased proportion of non-farm sector in rural employment, is an indicator of development since the proportion of non-farm sector in total employment (RNFE) increases as an economy develops. The states at different levels of rural diversification have been listed in Table 10.1. The RNFE is less than 20 per cent in some states, while in many other states, it is more than 35 per cent.

Table 10.1: States with different levels of rural diversification in the year 2004-05

RNFE (%)	States
Less than 20 per cent	Arunachal Pradesh, Chhatishgarh, Karnataka, Madhya Pradesh, Meghalaya, Mizoram
20.0-24.9 per cent	Bihar, Gujarat, Maharashtra, Nagaland, Uttaranchal
25.0-29.9 per cent	Andhra Pradesh, Assam, Rajasthan, Uttar Pradesh, India
30.0-34.9 per cent	Himachal Pradesh, Jharkhand, Manipur, Orissa, Punjab, Tamil Nadu
More than 35 per cent	Delhi, Goa, Haryana, Jammu and Kashmir, Kerala, Sikkim, Tripura, West Bengal

The National Sample Survey Organization's (NSSO) quinquennial survey of 2004-05 on employment shows that 27.3 per cent of rural workforce is in non-farm sector on the basis of the usual status of employment. Manufacturing leads the rural non-farm sector with 8.1 per cent of rural workers. The other industries (with shares in parentheses) are: trade—hotels and restaurants (6.1), construction (4.9), community social and personal services (4.5), finance-insurance-real estate and business services (0.5), mining (0.5) and utilities (0.02). Statistics at the state level have shown significant variations in the share of industries. As compared to previous quinquennial survey results on employment, the industry that has registered maximum increase in participation of rural workforce over the previous period is 'construction', followed by trade in hotels and restaurants (THR), manufacturing, and transport-storage and communication (TSC). Also, a perceptible decline has been observed in community-social-personal services (CSP) and mining industries during the period.

The above discussion indicates a pattern of occupational diversification in rural India. To assess potential of different industries, a comparative account of employment and income in major industries during the period 1983-94 and 1994-2000 has been presented in Table 10.2. The Table 10.2 shows the annual compound growth rate (ACGR) of employment in the rural and urban sector separately. The income data were obtained from the Central Statistical Organisation (CSO); however, income data for industries in rural and urban sectors were not available separately.

Employment in agriculture was largely rural-based and it was astonishing to find that the growth in rural employment in agriculture was abysmally low (0.06 per cent) and insignificant during the 1990s (Table 10.2). The corresponding growth was moderate and significant (1.13 per cent) during the 1980s. It was interesting to find that the growth in agricultural income during the 1990s was more or less the same as in the 1980s. These trends suggested '*job-less growth in agriculture*' during the 1990s (Jha, 2006b). The annual compound growth rate (ACGR) of employment in the non-agricultural sector, unlike for agriculture, has been found positive and significant during the 1990s; for both, rural and urban sectors. The ACGR for employment in the non-agricultural sector during 1994-2000 was less than the previous reference period, 1983-1994.

The non-agriculture industrial categories where employment growth during the 1990s was positive and also higher than that in the previous reference period, were manufacturing, construction, trade, transport and business services. Employment growth in these industries was slightly different at the levels of rural and urban sectors. In the rural sector, construction, transport and business services, recorded a higher growth during the 1990s than during 1980s. Interestingly, employment intensity in these industries also increased during the 1990s. However, as these industries accounted for only 11 per cent of the rural employment, employment intensity in the non-agricultural sector could not increase during the 1990s. Manufacturing, one of the most important industries in the non-farm sector, recorded a positive employment growth during the 1990s, but the rate of growth was lower in the rural than urban sector.

Table 10.2: Growth in Employment and Income for Selected Industries/Industry-Groups During 1980s and 1990s

Industries/industry groups with broad industry divisions (NIC 1998) in parentheses	*ACGR in employment*				*ACGR in income*		*Employment elasticity*	
	1983-94		*1994-00*		*1983-94*	*1994-2000*	*1983-94*	*1994-2000*
	Rural	*Urb*	*Rura*	*Urb*				
Agriculture (01-05)	1.13	2.33	0.06	-1.58	1.22	1.24	0.95	0.01
Mining, etc. (10-14)	1.47	1.47	0.27	-1.56	2.61	2.21	0.56	-0.04
Manufacturing (15-37)	0.89	0.85	0.84	1.32	2.52	3.10	0.34	0.35
Utilities (40-41)	0.41	0.67	-0.08	-1.22	3.51	2.92	0.16	-0.26
Construction (45)	1.03	3.11	2.28	2.61	2.10	2.67	0.82	0.89
Trade+hotels and restaurants (50-55)	1.67	1.88	1.22	4.31	2.36	3.81	0.76	0.82
Transport+storage+communication (60-64)	1.16	1.01	2.93	1.92	2.57	3.89	0.43	0.59
Finance + insurance + real estate + business services (65-74)	1.18	1.62	1.90	2.72	4.18	3.48	0.36	0.73
Com+social+personal services (CSP) (75-99)	0.66	1.93	-0.63	-2.40	2.40	3.37	0.59	-0.47
Non-agriculture/farm	1.03	1.57	0.91	1.24	2.70	3.39	0.48	0.32
Total	1.11	1.64	0.26	0.99	2.19	2.79	0.54	0.16

Note: These estimates have been worked out with the 'Current Daily Status' (CDS) figures of employment from the NSSO and income figures from the CSO, New Delhi.

During 1990s, employment growth was negative in mining and quarrying, utilities and community services. The decline of employment in these industries was due to the efforts of downsizing the public sector. Interestingly, income growth in the CSP services was positive. Income in these industries was basically the salaries of employees and with the implementation of the Fifth Pay Commission recommendations towards the late-1990s; income in these industries had increased. In the mining sector, decline in employment could also have been on account of the strict environmental regulations and increased focus on clean technologies.

The above observations suggest that in the 1990s growth of employment in agriculture was *plateaued*; growth in manufacturing also decelerated in the rural sector; and employment diversification in the rural sector was propelled by construction, trade, transport and business services; this, however, accounted for only 11 per cent of the rural workforce in the country. Therefore, agriculture and manufacturing hold the key for a sustained growth of rural livelihood in the country

Policy Options for Increasing Rural Diversification

Since employment and output in industry are highly correlated, it is difficult to separate industrial development policies, from those policies that are targeted towards creation of rural employment in industry. This separation has become even more difficult with the increased importance of cost efficiency in the liberalizing world. Government policies related to different sectors of economy have, therefore, been discussed and policy prescriptions for a robust growth in these sectors have been elaborated subsequently.

The other important policy options covered in this section are related to infrastructure and institutional aspects. Also, the issues pertaining to the direct employment—generation programmes (EGP) that have become almost a part of our development planning since past few decades have been elaborated. The empirical evidence to suggest that "trickle-down" effect of growth was not working effectively and the incidence of employment was on the rise, has increased the importance of such programmes.

Sectoral Growth Policies: As per the CSO classification, there are nine industries in which growth is important for the development—induced rural diversification in the country; however, the present study has observed that growth in agriculture and manufacturing only was important for a remunerative and broad-based growth of rural employment. Growth in these industries would trigger employment growth in other industries. Considering greater emphasis of other chapters of this volume on agriculture, the present subsection has discussed policies related to agriculture in brief, whereas manufacturing sector has been discussed in detail.

Agriculture: Government policies related to agriculture during the planned development of the country have passed through at least three distinct phases. The first shift in policy was evident in the mid-1960s with increased importance of self-sufficiency; the second line of demarcation surfaced with the opening up of trade in the early-1990s. It is really difficult to separate out employment-related policy in agriculture from the sectoral policy. Any government policy which increases cropped area and crop productivity would increase employment and wages in agriculture. Some suggestions for increasing cropped area and productivity in agriculture involve technological innovations, infrastructure development, rationalization of farm input prices, besides other issues discussed in detail in the study by Jha (2006a).

There is also a scope in increasing vertical integration of farm and firm. The less-exploited options in the form of apiculture, sericulture, and rearing of birds and small ruminants need to be encouraged, specially on small farms. This will increase the utilization of family labour, especially of women and can make the small farm more viable.

Rural Manufacturing: The rural sector has a definite advantage over its urban counterpart in manufacturing certain groups of commodities. The first set of products are those which utilize local resources and are semi-processed. These are not very scale-intensive, and examples in this category include honey and organic foods.

The second set of products are also based on local resources, are largely processed, and are not scale neutral. Technologies that are capital-intensive in nature often play a significant role in this category. Agro-processing and ancillary units near urban centres are examples in this category.

The third set of products consist of unmanufactured or semi-manufactured items; these are also labour-intensive examples in this category are traditional crafts by rural artisans. The Government of India has created specific institutions for the promotion of these industries, among which specific mention may be made of the Khadi and Village Industries Commission (KVIC). Certain government schemes specifically targeted towards encouraging rural manufacturing include the reservation for small-scale industry (SSI) and cluster programme.

The Khadi and Handloom Boards at the state level and innumerable institutions and cooperative societies at the disaggregate level were created for the development of khadi and village industries in the country. The khadi and village industries are launched to promote local resource-based products and traditional crafts in the rural areas. The KVICs promote rural entrepreneurship. The KVI products attract fiscal concessions; yet these products are often not cost-efficient. Inefficiencies in the KVI-system are often cited as reasons for this. The KVIC has also introduced a franchise scheme for its products. To improve the quality of KVI products, the KVIC in recent years has launched some brands such as 'Sarvodaya' for fast moving capital goods like toilet soaps, pickles, honey; 'Khadi' for the upmarket; essential oils and herbal products; and 'Desi Aahar' as organic food, cereals, and spices.

The government carved out SSI in its industrial policy resolution; this policy resolution also created several institutions to promote the small-scale sector (Jha, 2006a). A significant proportion of small industries are in the rural sector. Ttherefore, a robust growth in the SSIs is important for the growth of rural manufacturing. The government has undertaken certain steps in recent years to promote growth in these industries; thus, investment ceilings for the small-scale industries have been hiked to Rs 10 million, for selected items, this hike has been to the extent of Rs 50 million. The government has also attempted to revive the sector by infusing credit through SSI specialized bank branches, small and medium enterprise fund under SIDBI, laghu udyami credit card scheme, etc. However, certain problems specific to the SSIs such as prevalence of inspector-raj beyond the turnover of Rs. 10 million, adhocism and arbitrariness in de-reservation, high cost in obtaining basic goods and services because of prevalence of public sector monopolies in these industries, arbitrariness in import tariff charges for goods already reserved in the SSI or for similar competing goods in the SSI, etc. still remain unresolved. In an opening economy, the very concept of reserving industries is not tenable; therefore, the uncertainty associated with the reservation of SSIs must be finished with some categorical position regarding this. The performance of small and tiny industries also depends on the economic performance of some public sector monopolies that provide basic goods and services. Many of these monopolies have unfortunately less regard for cost efficiency, while the unit cost of production is becoming important in an open economy.

In a globalizing world when technology, cost and quality have become important, rural industrialization cannot rest solely with the KVI and SSIs. Increased private participation is desired to achieve a robust growth in rural manufacturing. Favourable infrastructure, largely under the public domain needs expansion. International experiences suggest the creation of industrial clusters in the rural vicinity as an effective measure of rural industrialization process. The union government had identified 60 industrial clusters in July 2003 for the focused development by incorporating their credit requirements in the state credit plan. The KVIC with the help of SIDBI and NABARD under the auspices of Ministry of Agriculture and Rural Industries, has also been trying to implement a National Policy for Agriculture and Rural Industries (NPRI). This essentially consists of technological advancement and skill up-gradation for the effective development of industrial clusters at the district level. This scheme aims to promote participation of the private entrepreneurs and NGOs. The Ministry of Food Processing Industry has also set up 'food parks' in different parts of the country. It is to provide capital-intensive common facilities such as cold storage, warehouses, quality control laboratories, effluent treatment plants, etc. to the adjoining processing units. The public sector units or corporate or even cooperatives are eligible for grants up to Rs 4 crore for the creation of such facilities. So far, 20 food parks have been sanctioned; however, their actual implementation is not known.

In spite of these efforts the performance of rural manufacturing in the 1990s has not been satisfactory. The reasons can be numerous; policy impediments for specific rural industry, burgeoning gap between rural and urban infrastructure, decline in demand for products manufactured in the rural sector. There is always possibility of creating suitable conditions and institutions for rural manufacture.

The performance of KVIC, which plays an important role in production and marketing of unmanufactured or semi-manufactured products, has been far-from satisfactory. To improve the efficiency of KVIC, there are suggestions to convert it into a promotional and developmental agency rather than a financial agency. The KVIC should provide technical support to the KVI units; including product process research and market research/strategies for the KVI-products. Since the village resource-based products have a niche in the international market, the KVIs need to do some aggressive marketing for such products. While adequate checks are needed for irregular release of export incentives, *ad-hocism* in the provisioning of rebate to the khadi sector may be abolished and a medium-term strategy on rebate must be chalked out so that production and planning are not disjointed.

There may be scope to involve the private sector and NGOs in the development of khadi and village industries for optimal utilization of resources and harvesting opportunities in a liberalizing world. Though the rural sector has distinct advantages in agro-processing, many large-scale processing units have not emerged in the rural region in many parts of the country. Organized growth of the processing industry also requires emphasis on the post-harvest infrastructure development. There is a scope for increasing private participation in the development of post-harvest infrastructures such as silos and warehouses, cold storage facilities and air-conditioned transport. The emergence of contract farming in some parts of the country is also one of the ways by which the benefits of agro-processing can be reaped by the farmers. Among other things, a wider dissemination of contract farming requires promotion of formal relations between the growers and industry.

The rural sector in India has a large number of artisans. In a liberalizing world, when distance is shrinking there may be a latent demand for skill-intensive, artisan-like work in a distant market; tapping of which requires the help of market professionals. Rural artisans as per their abilities and emerging opportunities may be trained on a selective basis; while care needs to be taken to ensure that this training must be integrated with the production and marketing of these products. The NGOs and producers' associations may be encouraged to undertake the job of training, producing and marketing the rural artisan's products. NGOs may encourage the formation of producers' SHGs to share the benefits of lucrative prices in distant markets; a good example of which has been presented by the *lijjat papad.* The formation of clusters would also help artisans in removing many of their size-and scale-specific bottlenecks.

The subject of rural industrialization is related to many government departments/ministries; these include, Ministry of Industries, Department of Agriculture and Rural Industries, Ministry of Rural Development, Department of Food Processing Industries. Similarly, public institutions created for rural industrialization or entrusted with the job of rural industrialization are also numerous; there is a possibility that a significant part of government expenditure is being spent on the maintenance of too many institutions; lack of proper coordination among these institutions also leads to tardy progress of rural industrialization in the country.

The above discussions show that the rural sector has the advantage in certain types of manufacturing activities. Depending on the typology, these manufacturing activities require different kinds of institutional support. In labour-intensive, capital-light, local resource-based manufacturing activities, improvement in the functioning of KVIC, producer-based small SHGs are desired; whereas, capital-intensive rural manufacturing activities require common-facility centres and a facility for industry-agglomeration. As rural manufacturing is the concern of many government departments, poor coordination between these departments often results in duplication of some programmes/efforts, while tardy a limited progress occurs in other programmes related to manufacturing activities.

Other Sectors: The previous section has shown that employment has declined in utilities and community services. Since these industries fall largely under the public domain and government is reducing its staff strength, employment under these categories has declined. However, demand for these services has not declined; in fact, with the growth of population and income and shrinkage of common resources, demand for these services has increased. There is enormous potential for private participation in the delivery of utilities and community services and thereby increasing rural employment in the country.

Though agriculture and rural manufacturing bear the onus of employment growth in the rural sector, several other industries like construction, trade, transport and business services have emerged as important ones in recent years. Employment growth in these industries depends on factors, such as infrastructure, per capita income, population density, etc. that are in turn influenced by the government policies to a large extent.

Infrastructure and Environment: Certain government policies transgress across industries/sectors; these include credit and infrastructure—related policies. Traditionally, the state is perceived as the provider of infrastructure and public institutions as the creator of rural technology. This perception is, however, changing with the emergence of new institutional alternatives. The sustenance of such institutions also requires government support, which may be in the form of suitable legislative environment. Another set of government policies increases capabilities of workers by providing better education and health facilities. Still another set of government policies attempts to protect vulnerable workers from such contingencies as illness, accident, untimely death of bread-earner, old age and unemployment; these are generally referred to as social security for workers. Some of these issues are illustrated below.

Technology and Skill Formation

Technology generation and dissemination, especially of interest to the rural sector is largely in the public domain. There is a general feeling that public-funded research and extension system in India is less responsive to the need of the people. The response of the farmers or target groups may be assessed with their willingness-to-pay for the research and extension services. With the user's payment principle, the research and extension system may emerge as more accountable and self-sustaining in the long-run. In this regard, the government has made some progress in terms of the agri-clinics and there are also suggestions for establishing 'village knowledge centres'.

The information technology (IT)-enabled knowledge dissemination centre can potentially provide a viable solution to individual farmer's problems at a distant isolated place. Public-private partnership, wherein technology generation would largely be in the public domain and its extension would partly be with the private domain or voluntary organizations shall be instrumental in taking the technology to rural doorstep. Some voluntary organizations are documenting available technologies and disseminating them in the rural areas. Such efforts, however, need to be consolidated and further replicated in a large part of the country.

The existing training infrastructure that has some bearing for the rural sector, such as the District Industry Centre (DIC), Industrial Training Institutes (ITIs), rural poly-techniques is in bad shape. These institutions must be revitalized and made relevant to the local needs. These rural institutes should also identify newer trades for training taking note of the resource endowment of the region and also the emerging opportunities in the liberalized/globalized context. Such trainings should be coordinated by local institutions, like DIC, District Rural Development Agency (DRDA), and others.

Finance and Infrastructure

In spite of so-much emphasis on institutional credit for farm and non-farm sectors, these sectors remain capital-starved. Financial institutions, on the other hand, often miss their targets for priority sector lending. The supply-demand mis-match in rural credit is often on account of lack of sufficient collaterals. The concept of community collateral has emerged as important alternative in recent years. A self-help-group (SHG) is one mode of arranging such community collaterals. The SHGs in addition to serving the needs of individual small-scale finances, may also resolve some specific problems associated with small-scale production of the non-farm sector. The performance of SHGs has, however, not been uniform across the country. Among reasons for this, the skewed distribution of non-banking financial institutions (NBFI), which is instrumental in the disbursement of credit to SHGs in the states, is the most important one. To encourage the activity and distribution of NBFI across the country, the credit limits of NBFI may be reduced. The Small Industries Development Bank of India (SIDBI) also funds the SHGs through NBFIs. The required credit limit for the NBFI is too high (Rs 2 crore).

There are suggestions to involve regional rural banks (RRBs) in disbursing credit through SHGs, since RRBs are relatively better distributed in the country. As there are already discussions about the ways tc increase the viability of RRBs, this additional work will spread their portfolios and may help in making these units more viable. A uniform spread of SHGs in addition to other factors requires many credible NGOs in a large part of the country.

A part of the need for credit will be solved if the community collateral and borrowing for viable industry and, or product-specific infrastructural project is allowed on an extended scale. To overcome some other problems related to collateral, industry associations and cluster-level units, may be encouraged to form a mutual credit guarantee fund. Service sector units such as trading houses/agencies which assist in marketing or brand buildings of rural products, may be given due priority by the banks and financial institutions.

The kind of infrastructures required for a specific industry or a product-group varies. Some industry-specific infrastructures like, cool van, quality-control laboratories for agro-processing, etc. may be initiated by the private parties or producers' and traders' associations. They definitely require a favourable incentive structure. The government may focus on the development of basic infrastructure like road, power and communications, since in the rural areas such basic infrastructure would largely remain in the public domain. In this context, there are suggestions that the Rural Infrastructure Development Fund (RIDF) may also be used more liberally in making investments in projects other than irrigation. The existing RIDF disbursements across states are highly skewed so that for generally prosperous states, the share in total disbursement is higher. A better regional distribution requires soft conditions for disbursement of rural credit under RIDF. Recently, the government has launched the programme of 'Bharat Nirman' precisely to strengthen the rural infrastructure in the country.

Social Security and Labour Welfare

Experience shows that in a labour surplus economy as that of India, growth in an industry is often not sufficient for the sustained growth of rural livelihood in that industry. Therefore, the government mediates in the rural labour market, in the form of assuring minimum wages or provisioning of satisfactory social security.

However, most of the social security and labour welfare policies in India cater to the organized sector, although it is the unorganized sector which provides a bulk of the employment to the rural people. There is a general feeling that workers in the organized sector are over-protected, while their counterparts in the unorganized sector lack even the minimum social security provisions. The nature of rural employment is often cited as the possible reason for such apathy. Employment in the non-farm sector is often seasonal, earnings are irregular and low and in many cases, the employer-employee relationship does not exist. Even if employer-employee relationship exists in some establishments, these are often not registered with the state governments.

Minimum wages is an important instrument for the provisioning of labour welfare, specially in the rural areas, though instances of violations of minimum wages are numerous. The legal limitations, definitional constraints and overburdened courts are often cited as reasons for the increasing violations of minimum wages. In the unorganized sector, where workers are regular and an employee-employer relationship exists; the provisions of minimum wages and social security to workers only require that it may be made obligatory on the employer. This of course requires identification of such establishments; the second NCL suggests enactment of the Small Entrepreneurs (Employment Relations) Bill to cover all establishments employing up to 19 workers and protection to all aspects of workers including wages, social security, safety and health. The NCL also proposes an umbrella legislation to ensure minimum wages to workers in the unorganized sector; a worker for this purpose is defined as one who is registered with a government agency and would permit the administrative body to decide the matter in case of dispute.

The absence of the employer-employee relationship in certain rural enterprises, require some innovative schemes suitable for particular micro-settings. Few state governments have attempted to create welfare fund for the target groups of producers like, *bidi* workers, by collecting 'cess' from the consumers of finished products. Certain state governments in association with non-governmental organisations (NGOs) have introduced social security schemes for specific target groups of workers; some of the successful schemes out of these experimentations need to be replicated throughout the country.

The government has recently introduced the Social Security Group Insurance Scheme for the unorganized workers in coordination with Life Insurance Corporation of India. Often, the reach of government social security schemes is limited because of poor literacy and awareness of rural workers. Some of the anomalies specific to rural workers can be reduced by the formation of Self-Help-Groups (SHGs) of workers employed in similar activities. The SHGs so formed can participate in certain welfare schemes of the government. The SHGs with the help of NGOs can interface with the government agency in a more effective way.

In India, the expenditure on social security is also low (less than 2 per cent) as compared to spent by many similar countries like, Sri Lanka (4.7 per cent) and China (3.6). Further, a large part of this expenditure is being incurred for the organized sector. Recently, National Committee for Enterprises in the Unorganized Sector under the chairmanship of Prof. Arjun Sengupta has drafted a scheme to provide the benefits of health insurance, life insurance and old-age security to the entire unorganized workforce of the country. An alarming situation on account of social security for the unorganized workers suggests that the committee report may be adopted.

Direct Employment Generation Programmes

The employment generating programmes largely fall under two broad categories; self-employment—generating and wage-based employment generating programmes. The first set of employment generating programmes (EGP) attempts to remove chronic unemployment by providing economic assets to the beneficiary, while the second group of programmes provides supplementary employment to stave-off seasonal unemployment. These employment—generating programmes have been part of development planning for more than three decades now. Though these programme have been modified time and again to improve their efficacy; the scale of employment provided by these programmes has been inadequate. In the light of deteriorating unemployment scenario, the government has started National Rural Employment Guarantee Programme, which legally guarantees employment to one person in every poor household for a minimum of 100 days on asset creating public works. However, initiatives of such a large magnitude have other implications as well. It is generally felt that increase of expenditure in such programmes is often associated with a decrease of public expenditures in the social sectors which is important for a long-term growth of the rural sector (Mahendra Dev, 2000).

Though there have been efforts in recent years to unify many of the wage-based EGPs, there is a possibility of further consolidating employment generation and developmental programmes. Examples of some labour-intensive rural works programmes (RWPs) undertaken by different government departments are: soil conservation, watershed development, and construction of schools and *pucca* roads. There is a feeling that if works under these departments are pooled and coordinated at the district level, it would leave sufficient funds for broadening the wage-based employment programmes and for creating durable community assets simultaneously in the rural sector.

With decentralization under PRI, the unification of various labour-intensive rural works programmes may not remain a far-fetched idea. It probably requires formulation of district level plans and proper coordination between the elected body, bureaucracy, and professionals/subject specialists at the district level. The district-level plan may take some more time to realize. Meanwhile, a clearing house at the district level may be setup for the sharing of information by various agencies *administering* the employment generation programme so as to avoid duplication of beneficiaries and the creation of durable assets in the programme.

Conclusions

The study has found that rural manufacturing is the most important industrial category in the non-farm sector, but employment growth in it has decelerated during the recent decades. There are also evidences depicting shifting of manufacturing activities away from the rural sector to urban sector in the country. Construction, trade, transport and business services have spearheaded growth of non-farm employment in the rural sector. As these industries together account for only 11 per cent of rural workforce in the country and employment growth in them is not autonomous; therefore, there are fewer chances that these industries would remain the driver of rural employment growth in the country. Strategies to increase rural non-farm employment must lay greater stress on agriculture and manufacturing; though, growth in other sectors does remains important.

A review performance of the public institutions has suggested sufficient scope of improvement. Stakeholder's participation in the management of these institutions and a levy of user's charges for availing of the services of public institutions are some suggestions to make these institutions more accountable. Institutions and technology that cater to small and scattered production units are desired. Certain innovative institutions have emerged in select parts of the country; that need to be replicated over a wider area. In addition to the emphasis on important sectors of economy, the rural works programmes remain important for at least three reasons: increased marginalisation of agricultural land, increased seasonality of employment, and the importance of public goods in the rural sector. Few anomalies in direct employment generation programmes can be identified with concurrent evaluation of such programmes. With decentralization and effective participation of *panchayti raj* institution, some of the frequently cited weakness of employment generating programmes will be eliminated.

Note

1. A subdued domestic price of agricultural commodities during 1999-2002 was often cited as an important reason for farmers distress during the period; however, with upswing in international prices, domestic prices of many commodities have increased, so distress on that account should also have reduced. This is an example of periodic reasons of farmers' distress.

References

Jha, B. (2006a), *Rural Non-farm Employment in India: Macro-trends, Micro-evidences and Policy Options*, IEG Working Paper Series No. E/272/2006. Institute of Economic Growth, Delhi.

Jha, B. (2006b), *Employment, Wages and Productivity in Indian Agriculture*, IEG Working Paper Series No. E/266/2006. Institute of Economic Growth, Delhi.

Jha, B. (2007), Is the role of agriculture in rural diversification decreasing? *Indian Journal of Labour Economics*, 50(4): 633-642.

Mahendra Dev, S. (2000), Economic liberalization and employment in South Asia, *Discussion Paper on Development Policy* No. 29, Centre for Development Research (ZEF), University of Bonn, Germany.

NSSO (National Sample Survey Organisation) (1990), *Results of the Fourth Quinnquennial Survey on Employment and Unemployment (All India), NSS Forty-third Round (July 1987-June 1988)*, NSS Report No. 409, New Delhi.

NSSO (1997), *Employment and Unemployment in India, 1993-94, NSS Fiftieth Round (July 1993-June 1994)*, NSS Report No. 409, New Delhi.

NSSO (2001), *Employment and Unemployment Situation in India, 1999-2000, NSS Fifty-fifth Round (July 1999-June 2000)*, NSS Report No. 458, Key Results, Part-I, Part-II, New Delhi.

NSSO (2007), *Employment and Unemployment Situation in India, 2004-05, NSS Sixty-first Round (July 1999-June 2000)*, NSS Report No. 515, (summary of findings), New Delhi.

Rani, Uma and Jeemol Unni (2004), Unorganised and organised manufacturing in India–Potential for employment generating growth, *Economic and Political Weekly*, 49: 4568-4580.

Section IV

STRATEGIES TO HARNESS OPPORTUNITIES

While opportunities exist to alleviate agrarian distress through a combination of farm and non-farm options, what is essentially required are the right kind of strategies to harness these opportunities. The approach to address agrarian distress should essentially encompass the measures for prudent financial management, mitigation of risks in agriculture, widening the outreach of support services and social networks and rational input-use in farming. There exists a wide range of short-, medium- and long-term strategies that stem from these approaches to deal with the agrarian crisis and has put the agriculture sector on the track of faster growth and prosperity. Section IV focuses on three important strategies; one, reorientation of the agricultural research and extension system; two, revamping of the institutional credit delivery system and three, development of the agro-processing industry.

The National Agricultural Research System (NARS) has an extensive network of institutions aimed at providing R&D, HRD and technology dissemination services to the agricultural sector. The System has made exemplary contribution in ushering an era of self-sufficiency in foodgrain production and has been the leader of agricultural progress in the country. However, in the changing global and domestic economic environment, there is a need to chalk out a clearly stated strategy that rationally prioritizes the research agenda. In the context of general slowdown of agricultural growth and increasing agrarian distress in some particular regions, a major paradigm shift is needed to transform the present commodity-based research to a systems approach. Since farm-level problems are region-specific, convergence is needed between R&D agencies within the regions so as to bring region-specificity in technologies and their time-bound assessment. Reorientation of the agricultural research and extension system is ardently required to establish a research-development-technology transfer continuum involving all stakeholders.

Since mounting indebtedness of farmers has been viewed as one of the prime reasons for farmers' suicides, the government has taken several measures to ease out the situation, including the 'loan waiver scheme'. These measures can at best be considered short-term steps to address the problem. The long-term strategies to make credit as an instrument for revitalizing the agricultural sector should aim at financial inclusion in terms of innovative products and services for increasing access to institutional credit. The study on rural credit delivery system presents a detailed account of the present and future strategies for making the system more responsive to the needs of customers.

The development in agro-processing is essential to sustain the demand for high-value commodities. The food processing industry provides a direct market for the farmer on the demand side of the food chain. Processing of plant and animal output into the value-added products enables fetching of remunerative prices and provides significant employment generation across the supply chain in production of raw materials, manufacturing, storage and distribution of finished products to the consumers. On account of convenience and food safety, the policy strategy for linking farmers with domestic and international consumers; create enabling environment for reaping the effective benefits of diversification towards high-value commodities. Public-private partnership models to strengthen the food processing industry have also been discussed at length in the last Chapter of the section.

11

Role of Agricultural R&D Policy in Managing Agrarian Crisis

Sant Kumar, Suresh Pal and Rashi Mittal

Introduction

The major cause of agrarian crisis, as brought out in the appraisal of 10th Plan document, is loss of dynamism in the agriculture sector. More specifically, agriculture sector is facing the problems of declining factor productivity and supply-side constraints that call for efforts in refining and developing the technology. In the times to come, the potential adverse impact of climate change could further compound the agrarian crisis in the absence of efficient management of natural resources like land and water. It is a real technology and policy challenge to cope up with the emerging crisis.

Research and development (R&D) has potential to offer long-term solutions to the problems of agriculture sector. Consistent R&D efforts in agriculture have helped in up-scaling the potential and have provided options to derive same level of benefits or even higher benefits at lower cost per unit of output. In the current context, though technology alone is not able to provide the complete solution for managing the crisis of the sector, it is capable in offering better possible options.

Over the years, achievements in R&D and increased awareness have resulted in adoption of recommended practices and technologies by the farmers. However, the extent of dissemination of the technologies has not been adequate. Among the various reasons for low adoption of technologies at the field level, the weak linkage between research and extension is a serious impediment. It is believed that research system in India is relatively strong in technology generation than supplying it appropriately to the end-users. Educating the farmers about improved technology and ensuring its continuous access and adoption is a major challenge for the effective deployment of the agricultural R&D system.

This Chapter gives a brief overview of the existing set-up of agricultural R&D in India, emphasizes the need for re-orienting the financial and manpower allocations in research and discusses the emerging need of new approach to technology transfer in the context of the agrarian crisis.

Organizational Set-up of Agricultural R&D in India

Presently, agricultural R&D in India is being managed under a three-tier system: (i) Indian Council of Agricultural Research (ICAR) at the central level, (ii) State Agricultural Universities (SAUs) at state level, and (iii) Private sector at both the sectoral and commodity levels. The agricultural research in India has historically been in the public domain and continues to be so (Table 11.1). The development of independent agricultural research and education (R&E) institutions can be traced back to the pre-independence period, which was further strengthened as a part of the 'new agricultural strategy' adopted during early-1960s. A need was felt to build up human resources (both scientific and technical) by promoting education, research and extension at the state level through the SAUs. The ICAR was reorganized and given a critical role in agricultural R&E. It initiated, funded, guided and managed a new phase of central initiatives embracing research, education and frontline extension.

The present ICAR-SAUs system is supported by All India Coordinated Research Project (AICRP) which is based on the principle of inter-disciplinary and inter-institutional collaborations. Starting with the first AICRP on Maize in 1957, the ICAR had 62 AICRPs during 2005-06 covering various disciplines and commodities, viz. soils, water, crops, horticulture, livestock, fisheries, home science, agricultural engineering, education, etc. AICRPs related to crops have delineated the operational area based on ecological conditions. This setup enables AICRPs to effective utilize natural resources, in man and material, and to solve the national, regional and local problems in a coordinated manner according to predetermined priorities and strategies.

In addition to the National Agricultural Research System (NARS), predominantly comprising of ICAR institutions and SAUs, there are institutions in central departments of agriculture, Council of Scientific and Industrial Research (CSIR), departments of science and technology, fertilizers and chemicals, commerce, etc. which also contribute to the national efforts through research institutes or research programmes (Jha and Kumar 2006).

Table 11.1: Status of Public Sector Agricultural Research System in India, 2005-06

Institutions	*Number*
Central	
National Institutes (Deemed Universities)	4
Central/Other Institutes	48
National Bureaux	5
Project Directorates	12
National Research Centres	32
All India Coordinated Research Projects	62
Central Agricultural University	1
Krishi Vigyan Kendras	500
Zonal Coordination Units	8
State	
State Agricultural Universities	39
Agricultural/Zonal Research Stations	343*

*Includes 126 zonal research stations.

Source: ICAR (2007), and Ghosh (1991)

Over the period of time, the research environment has undergone several changes. The private sector that was virtually non-existent in agricultural research about four decades ago, is now, playing a critical role in meeting R&D needs. The entry of private sector R&D in India started with liberalization of seed sector in mid-1980s and picked up momentum in late-1990s as incentive for private investment in R&D improved. But largely, private investment has increased in those activities where returns are less risky, quick and relatively higher. In few cases, private sectors have come forward to capture future markets and have invested huge amount of money in building R&D. The main reasons for private sector entering into agricultural R&D are enormous potential, opportunity and relaxation in entry provisions by the government. Private sector now accounts for 14 per cent of the total research funding in agriculture (Pal and Byerlee, 2003).

Allocation of Research Resources

Financial Investment in Agricultural R&D: Total government funding for agricultural R&E increased in real terms (at 1999 prices) from Rs 2.5 billion in 1961 to Rs 7.9 billion in 1980. It further increased to Rs 25 billion in 2000–a ten-fold increase over the past four decades. An increasing trend has been observed for both central and state funding. However, increasing centralization of public investment in R&E is a major concern. Data show that the share of states in total R&E investment has fallen from 69 per cent in 1971 to 50.5 per cent in 2000 (Figure 11.1). Local institutions have failed to emerge as major players and supporters of agricultural R&D. This is a major weakness which has not been well appreciated. The central system continues to press for and obtain incremental resources. States either do not bother or lack capacity to argue their case. This dynamics can only be understood in a political economy framework, which is beyond the scope of this Chapter.

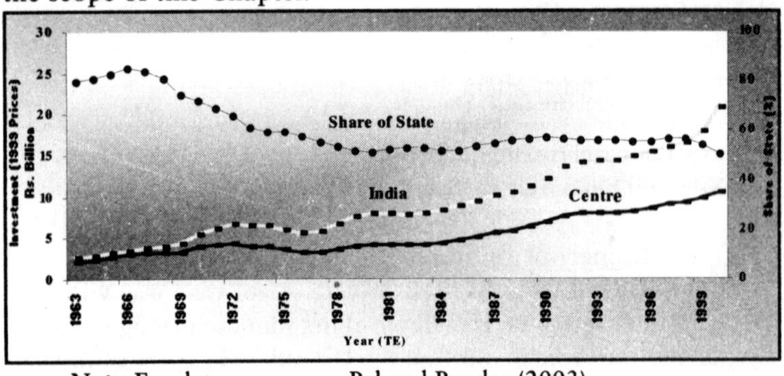

Note: For data source, *see* Pal and Byerlee (2003)

Figure 11.1: Public Expenditure on Agricultural Research and Education

Another way to look at the status of investment on agricultural R&E is to estimate investment intensity ratio of public R&E. The ratio is computed as investment on R&E as percentage of agricultural gross domestic product (agGDP). This ratio increased significantly during 1960s (0.20 per cent) and 1980 (0.36 per cent), but remained around 0.40 per cent during 1990s. The slowdown in investment intensity in relation to average intensity for all developing countries of 0.6 per cent and 1 per cent globally is a worrisome trend. The investment intensity ratio for Latin America and Africa are more than double of India. Further, there was a considerable variability in R&E intensity across different states of India during TE 2004 (Figure 11.2). Figure shows that with the exception of few states, commitment to investment in R&E is not strong and in some states (Uttar Pradesh, Orissa, West Bengal) the situation is more disturbing. The above analysis clearly shows a situation of under-investment in agricultural R&E in India.

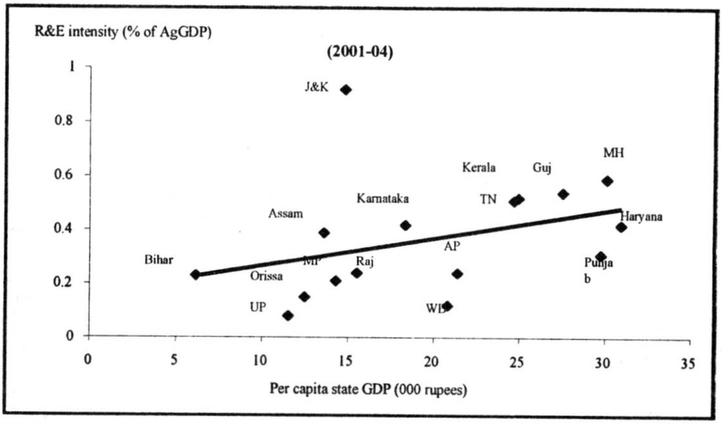

Figure 11.2: Per Capita State Domestic Product and Research and Education Intensity: 2001-04

Manpower Resources: Investment in agricultural R&D has also been observed on the basis of scientific manpower engaged in this sector. Scientific manpower is the most critical resource in R&D. As per the latest estimates, there are about 22,000 scientists engaged in agricultural R&D, of which nearly 96 per cent are in the public sector (Table 11.2). In-depth scrutiny has shown that ICAR-SAU system, which is the major entity of NARS, employs 83 per cent of scientists, the SAUs system alone accounts for nearly 62 per cent. Other public and private sectors' share 13 per cent and 4 per cent of total scientific manpower, respectively. Detailed analysis has shown that the state system has lost 24 per cent of scientists during 1992-2002 (Jha and Kumar, 2006). This is an important policy issue and needs attention.

Data has revealed that two-thirds of the scientists hold Doctorate degrees and rest are Masters. In ICAR-SAU system, the skill level is better (70-76 per cent). Analysis has also drawn attention towards the aging of scientists in the NARS. The average age (44 years) is creeping towards the level at which scientific productivity starts declining. More than 45 per cent of scientists are above the average age.

Table 11.2: Profile of Agricultural Scientists

Particulars	Number
Total scientists	21,869
(a)I CAR	4,539
(b) SAUs	13,633
Average age (years)	44.0
Ph.D. degree holders (per cent)	67.7

Source: Jha and Kumar (2006)

Scientists accord high priority to research (47.5 per cent), and those in the ICAR and private institutions spend relatively more time on R&D. In the SAUs and deemed universities under ICAR, teaching activity is the next in importance (26.6 per cent). Frontline extension gets the third rank in time allocation by scientists (18.8 per cent). Further, a large amount of scientists' time (11 per cent) is allocated to administration. This substantial time allocation to non-R&D activities (administration and management) is a worrying feature.

Pattern of Allocation across Commodity Groups: Research goals in agriculture have become more complex due to the fast changing situations and increasing supply-side constraints. In this regard, it is pertinent to use limited scientific resources efficiently. Efficiency of scientific resources has been observed in relation to the economic importance of different commodities measured in terms of their value of output (VOP). Figure 11.3 shows that allocation of resources to individual commodities appears to be in line with their relative economic importance. However, condiments and spices and to some extent, plantation crops are exceptions. The former appears under-researched and latter receives disproportionately high attention. Figure 3 also indicates that in some cases (cereals and vegetables) small adjustments are needed, while relatively large shifts are needed in oilseeds, plantations, fibres and pulses. But, before taking any readjustment decisions, policymakers need to consider other parameters like food security, nutrition, future demand, etc. Apart from these, this exercise is static in nature and to make any readjustment, parameters adopted in the exercise must be updated and their dimensions need to be broadened.

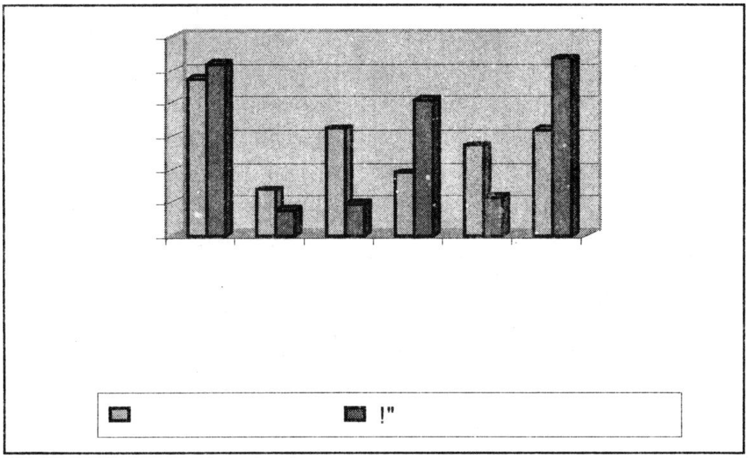

Figure 11.3: Allocation of Scientists by Commodity Groups

Regional Allocation Pattern: Rationality of research resources was also seen at the regional level. This issue is important in R&D planning due to strong interactions between technology with agroclimatic regions and socioeconomic conditions. Analysis suggests paying more attention in terms of allocation of manpower resources in the Gangetic Plains, and Semiarid Tropics in order to create a more rational allocation profile (Figure 11.4).

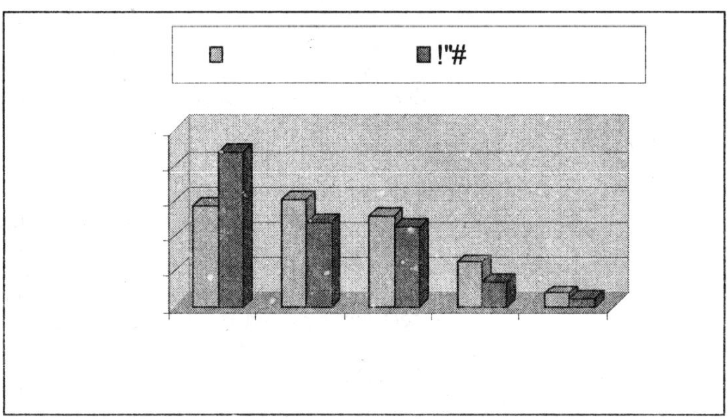

Figure 11.4: Allocation of Scientists by Regions

Status of Potential Agricultural Technologies

Over the years, ICAR institutes and SAUs have developed technologies for various farm activities. Studies have shown that improved technologies are capable of making farming operations easy, providing savings in resources, reducing post-harvest losses, increasing shelf-life and value and finally increasing food production. The stock of few potential technologies has been given in Table 11.3. The listed technologies cover various aspects, from improved crop

Table 11.3: Features of Potential Agricultural Technologies in India—contd.

S. No.	*Technology*	*Expected benefits*
1.	**Crop improvement**	
	Hybrid rice	Hybrid rice technology provides additional yields of 1.0-1.5 t/ha over popular check varie-ties. This technology is becoming popular in low-to-medium productivity areas of eastern India.
	Quality protein maize	It is the first single cross hybrid quality protein maize developed in the country. It possesses 40.7 per cent higher tryptophan than 'Vivek' maize Hybrid 9. It is suitable for growing in maize zones I and IV.
2.	**Improved crop management**	
	IPM	IPM technique is able to manage pest population of *Helicoverpa armigera*, and reduces pesticide use by 50 per cent. It reduces bollworm damage to 10 per cent, and provides additional cotton yield of 0.37 t/ha.
	IWM	Adopting good water management practices has increased rice yield (13 per cent in summer and 15 per cent in *kharif*) and water-use efficiency of 28.7 per cent and 52.8 per cent, respectively over the conventional practices of flooding.
	INM	Balanced use of N, P, K, S and Zn at the Recom-mended rates (120 kg N, 60 kg P_2O_5, 20 kg K_2O, 20 kg S and 5 kg Zn/ha) has increased wheat yield by 15-24 per cent in MP. Skipping the use of P and S has resulted in yield losses.
	Diversified farming System	Farming system comprising crops, horticulture, livestock and fisheries provides opportunities to profitability (3-4 times over monoculture). It also provides enough opportunity for increasing employ-ment, income generation and nutritional security.
3.	**Resource conservation**	
	Zero-tillage	Use of zero tillage has saved water by 11.3 per cent, and increased wheat yield (3.26 t/ha), against conventional sowing (3.23 t/ha) and has raised bed planting (3.21 t/ha).

Contd.

S. No.	*Technology*	*Expected benefits*
	Micro-irrigation	Ferti-irrigation through drip has improved quality and yield of banana. Fruit size was bigger (35-45 mm diameter) over non-drip (28-37 mm) for Harichal variety), with yield advantage of 273 per cent. The drip system improves overall profitability of the system.
	Pressurized irrigation system	Pressurized irrigation (sprinkler and drip) system helps in increasing yield by 40-50 per cent, and provides water savings of 30-70 per cent depending on the crop. Sprinkler irrigation is suitable for all crops (except rice and jute), while drip is more effective in horticultural crops, cotton and sugarcane.
4.	**Improved livestock methods**	
	Use of AI	Use of crystoscope in artificial insemination (AI) would improve conception in cattle and buffaloes by 20-25 per cent.
	Supplement of deficient minerals	Macro-and micro-nutrients deficiency in soil is affecting the productivity of livestock. Supplementing the deficient minerals in the diet of livestock would improve productivity by 10-15 per cent.
	Use of urea ammoniation technology	The quality of roughages fed to livestock is generally of poor quality arresting desired output. Proper use of urea ammoniation technology, would improve the nutritive value of roughages and thus production by 5-8 per cemt.
	Livestock health management	Malnutrition affects livestock as it does to human. Feeding of complete feed blocks sustain the production. These feed blocks can be transported to places of acute feed shortage.
	Diagnostic kits and vaccines	Diagnostic kits and vaccines are capable in controlling diseases and parasites like foot and mouth disease (FMD), infectious bovine rhinotracheitis (IBR), *peste des petitis ruminants* (PPR), bovine viral diarrhoea (BVD), bluetongue and avian influenza. Their use in livestock health management would save annually Rs 25,00-30,000 crore.
5.	**Improved machines *Production operations***	
	Land leveler	Use of leveller reduces water use by 20-30 per cent, increases crop yield by 10-15 per cent, with the land levelling cost of Rs 1500-2000/ha.
	Rotavator time (30-35	Rotavator has multiple use and saves per cent), water (30 per cent), and cost of operation (20-25 per cent) as compared to tillage by cultivator and harrow.
	Raised bed former	This equipment is useful for wheat after soybean, maize or cotton. The benefits in saving cost of operation (20-30 per cent), seeds (25 per cent), fertilizer (25 per cent) and irrigation water (20-30 per cent).

Contd.

S. No.	Technology	Expected benefits
	Ridge seeder	The ridge seeder is most suitable for dryland farming and is suitable for planting ragi, gram and pearlmillet. Its use increased the yield by about 15 per cent, against conventional method of sowing.
	Sugarcane cutter planter	It is suitable for cutting and planting sugarcane setts and application of granular fertilizer in single operation. The labour requirement and time of operation are reduced by 78 per cent and 50 per cent, respectively.
	Potato planter	It is used for potato planting and also for interculture and earthing operation. Its field capacity is 0.4 ha/hour, and cost of operation is Rs 750/ha.
	Rotary weeder	It is suitable for weeding wide-spaced crops such as cotton, tapioca, sugarcane, grapes, cotton, arecanut, and other orchard and plantation crops. Its cost of operation is Rs 880/ha, which is 35-50 per cent less as compared to manual weeding.
	Post-harvest processing	
	Wheat straw combine	This machine recovers 55-60 per cent straw in addition to 75-100 kg grains/ha, resulting into an average net saving of Rs 1,250/ha.
	Rice mills	Use of rice shellers outturn rice 70-72 per cent, against 65-68 per cent from traditional huller. At country level, its use can pay additional 2.0 million tonnes rice bran which in turn will give 0.25 million tonnes high quality rice bran oil (RBO) fetching Rs 1,000 crore.
	Modern ginneries	Modernizing cotton ginnings using variable speed double roller (VSDR) saves 30 per cent energy and increase efficiency including competitiveness in lint production.
	Biomass energy plant	
	Biogas plants (solid waste)	The improved biogas plant requires 75-100 per cent less water, produces up to 30 per cent more gas, requires one-fourth space for slurry storage/drying and costs 10 per cent more than the common design. The feeding of substrate and handling of the digested slurry is easier.

Source: Compiled from ICAR/DARE reports

production to resource conservation and value addition. The full potential of the available technologies is yet to be tapped. Therefore, developing mechanisms and providing support for wider adoption of the technologies by the farmers will be instrumental in managing the agrarian crisis.

Drawing from the factors responsible for poor and partial adoption of technologies on farmers' field, first and foremost, there is a need to refine technology suited to local conditions and make it available to farmers. This emphasizes the requirement of higher investment in agricultural research and extension, together with more developmental outlay for the agriculture sector.

Emerging Need of New Approach to Technology Transfer

The new agricultural strategy adopted in mid-1960s worked well till early-1990s, with few exceptions. However, as agriculture becomes more and more knowledge-and capital-intensive due to the changing economic situation and pressure of environmental problems, farmers need continuous and updated information on meteorology, marketing and a number of other aspects directly or indirectly affecting agricultural and allied activities. In fact, there is a need to launch a literacy movement at the farm level covering all aspects like, information related to genetically modified farm products, legal aspects of IPR and Farmers' and Breeders' Rights, quality requirements under sanitary and phytosanitary measures and Codex Alimentarius standards, agricultural trade, etc. Presently, there is a gap between information needs of farm families and that provided by the traditional extension agencies.

The availability of timely, location-specific demand-driven and value-added information is very important for the growth of agriculture. Adequate knowledge on part of the technology providers about local agriculture and socioeconomic conditions including various farming methods and techniques, is equally important. Apart from these, rural women and men need access to healthcare information. Information on health status of livestock and poultry, on-farm and off-farm livelihood options and market-led entrepreneurship opportunities for the poor and the marginalized in rural India need attention. There is also a need of promoting functional literacy among the adult illiterate and making learning joyful and useful for the young through interactive pedagogic methodologies.

Experiences have shown that rural women and men can have access to information and communication technologies (ICT) with ease and confidence provided there is opportunity for learning by doing and the contents are dynamic and location specific. ICT would be more meaningful to rural families if generic information is converted into a location-specific one. If properly designed, ICT can help in converting scientific know-how now available into field level do-how. Bridging the digital divide helps in knowledge and skill empowerment, and also helps in bridging gender divide by enhancing the self-esteem and self-confidence of rural women.

Summing-up

Agricultural R&D has enormous potential in providing solutions to the persisting problems. India operates with a three-tier R&D system. The ICAR-SAU system is the major sect of NARS in technology development and its dissemination. Role of the growing private sector is limited to few crops and sectors. To meet the social goals and reaching to far flung locations, strong public R&D system is a must. Presently, ICAR-SAU system shares equal proportions in investment on R&E. State system lacks capacity of increasing investment in agricultural R&E, although agriculture is a state subject. The investment intensity in R&E is 0.4 per cent against 0.6 per cent in all the developing countries and 1 per cent globally. Further, a wide variability in R&E intensity across states of India is noticed. Apart from financial, human resources in SAUs have reduced by 24 per cent during 1992-2002. Aging of scientific manpower is another problem. These facts clearly show underinvestment in agricultural R&D. However, the efficiency of manpower resources is more or less in line with economic importance for which these are allocated. Efficiency of allocation in regional context has suggested some readjustments in favour of the Gangetic Plains and Semi-arid Tropics that are important from food-security point of view.

In the wake of globalization and liberalization of economies, the policies promoting agriculture are also changing fast. With the entry of private sector, the issues of providing technology and distribution of incentives have come at the central stage. These situations are beneficial for resourceful farmers and allow them to harness opportunities but the poor farmers are left behind. This important issue is becoming a big hindrance in adoption of potential technologies developed by the private companies.

Technology development process has moved from the method of plant introduction to hybridization and better plant type for various agro-climatic conditions and better plant type. The ICAR-SAUs system has developed various improved technologies over the years and these have exhibited good potential in increasing income and employment and making farming a profitable source of livelihood. There is a need to develop mechanism and provide support for their adoption.

The fast changing situation and policies have made agriculture complex and more knowledge—intensive. In the regimes of trade liberalization and IPRs, farmers need more education and recent information to avail benefits. The role of ICT seems more meaningful in providing generic information under the local situation and empowering farmers with scientific know-how and information on markets and prices.

References

Ghosh, S. P. (1991) *Agro-climatic Zone-specific Research: Indian Perspective under NARP*, Indian Council ofAgricultural Research, New Delhi.

ICAR. (2007a) *ICAR/DARE Annual Report 2006-07*. Indian Council of Agricultural Research, New Delhi.

ICAR (2007b) *State Specific Technological Interventions for Higher Agricultural Growth*. Indian Council ofAgricultural Research, New Delhi.

Jha, Dayanatha and Sant Kumar (2006) *Research Resource Allocation in Indian Agriculture*. Policy Paper 23, National Centre forAgricultural Economics and Policy Research, New Delhi.

Pal, Suresh and Derek Byerlee (2003) *The Funding and Organization of Agricultural Research in India: Evolution and Emerging Policy Issues*, Policy Paper 16, National Centre for Agricultural Economics and Policy Research, New Delhi.

12

Harnessing Potential of Food Processing Sector for Accelerated Agricultural Growth

Tushar Pandey and Venetia Vickers

Food processing covers a wide spectrum of products from a large number of sub-sectors comprising agriculture, horticulture, plantation, animal husbandry, dairy and fisheries. With rising incomes, favourable demographic transition and changing consumption patterns, growth of food processing industry in India has become inevitable. Moreover, Indian farm produce has unique aroma, flavour and taste and if processed, packaged and marketed properly, has the potential to capture the world market. Promotion of the food processing industry is, therefore, essential for the national economy, farmers and consumers.

This Chapter reviews the status of food processing industry in India, highlights the opportunities for and constraints in the development of this industry and finally discusses the strategies for its rapid growth

Present Status of Food Processing Industry

India's food processing sector, contributes 14 per cent to the manufacturing GDP, produces goods worth Rs 2.8 trillion and employs 13 million people. However, the Indian agri and food processing industry is still nascent; only 2 per cent of the fruits and vegetables and 15 per cent of the milk production are processed as compared to about 80 per cent in Malaysia and 30 per cent in Thailand.

Currently, the Indian food processing industry is basically export-oriented. Although domestic consumption of processed food is low, it is fast picking up with rising income levels and changing consumer preferences due to economic growth.

Unorganized, small players account for more that 70 per cent of the industry's output in volume and 50 per cent in value terms. Most of them operate locally, add little, if any, value to the products, and use outdated technologies. However, over the past few years, several large companies, both, Indian and foreign, have invested in the food-processing business in India, resulting in significant growth in this sector.

The food processing industry in India has undergone significant changes over the past six to seven years, in terms of types, variety, quality and presentation of products, which is mainly a result of liberalization that led to foreign direct investment in the processed food sector. Most of the food processing sectors have been brought under the liberal, transparent and investor-friendly FDI policy, which allows 100 per cent FDI.

However, the small-scale farming system in India, marketing problems, lack of grading and standards and poor distribution channels, among others, continue to pose problems for the processing industry to source the right type of raw materials and discourage more investments in the sector.

There being several sub-components of food processing, a brief overview of the main sub-components is important for getting a better insight into the status of the industry.

Grain-based products: About 75 per cent of India's wheat production is milled into wheat flour (*atta*) to make *rotis* or *chapattis* (unleavened flat bread), mostly in small *chakkis* (wheat grinding mills) in the unorganized sector. Branded atta is a relatively new segment in the market, developed to provide consumers a more hygienic quality, as compared to gunny bag packed atta. Annual production of branded *atta* is about 1 million tonnes, and is growing at the rate of 7-9 per cent annually.

Bakery Products: These constitute the largest segment of grain-based processed foods. Small and medium unorganized local players and a limited number of organized units dominate the industry. The grain-based snacks market, comprising extruded snacks and savouries, is estimated at around Rs 29 billion. Of this, the organized segment contributes only 15 per cent of sales. Breakfast cereal production in the organized sector is very small, and is mainly confined to corn flakes.

Horticulture: India is the second largest producer of fruits and vegetables in the world, accounting for 10 per cent of global F&V production. Fresh packed F&V, primarily an export-led business, is a large industry in India, and is growing rapidly. Indian yields are among the best in banana and grapes. However, they are extremely low in most other F&V.

India is the land of spices producing several varieties worth over Rs 3500 crore amounting to 25-30 per cent of world production, which is processed for value-addition and export. Processed spice products are mainly in the form of curry powder/mixture, mint products and spice oil and oleoresins.

Edible Oils: India is the world's fourth largest vegetable oil economy with 15000 oil mills, 60 solvent extraction units, 230 vanaspati plants and over 500 refineries, employing over one million persons. A large number of multinational companies have set up operations in India through either port based refineries, trading subsidies or brand acquisitions in Indian edible oil sector.

Meat and Poultry: Poultry production has also shown a steady increase along with a sharp rise in the availability of eggs and broilers. With increase in the availability of eggs in the country and demand of egg products in Europe, Japan and other countries, India has started exporting egg products like, whole egg powder, albumin powder, yolk powder, etc. Recently, several companies have launched value-added poultry products. Although production of meat and poultry products has increased, only about 1 per cent of the total meat is converted into value-added products such as sausages, ham, bacon, luncheon meat, kebabs, meatballs, etc.

Dairy: Milk processing is low at less than 15 per cent. Currently, processed dairy products in India comprise milk powder, condensed milk, cheese and malted milk foods. About 90 per cent of dairy processing is in the form of milk powder. The trends indicate a slow rate of growth in all the processed dairy products.

Fisheries: Processing of fish into canned and frozen products is carried out mainly for the export market. This sector has attracted a very high quantum of investment, to the tune of about Rs 30 billion from 2000 to 2004, of which 23 per cent is the foreign investment. This quantum has increased further over the past three years. Over the past one decade, the organized corporate sector is becoming increasingly involved in preservation, processing and export of coastal fish. However, it is interesting to note that processed sector of the fisheries is wholly export oriented. Processing of fisheries for domestic consumption is rare.

Non-alcoholic Beverages: India is the largest tea producer in the world. Tea production in Darjeeling, Nilgiris and Brahmaputra valley are considered as the best quality tea in the world. Tea processing includes withering, rolling, fermenting, drying, blending, packing and branding.

With an annual production of around 300,000 tonnes, India is a small but competitive producer of coffee. The instant coffee segment is entirely branded and packaged and caters mostly to the export market.

Opportunities for Growth

With the changing lifestyle of consumers and the rising disposable income of the growing middle-class, there is an increasing demand for convenient and hygienic foods. Large and exceedingly wealthier middle-class is creating growing demand for a wider variety of high quality processed foods. Demand is also boosted by the changing age profile (sixty-five million people are expected to enter 20-34 year age group by 2010) and increasing exposure to western-type products and lifestyles. The cultural barriers of preference for fresh food and eating at-home are smashing due to urbanization, nuclearization of families, rise in the number of working couples, rising costs of domestic help, media exposure and finally emergence of the global consumer. Another recent trend towards a healthier life-style has generated a niche market for diet, healthy, low-calorie, and non-fat food products.

All this is expected to increase demand for processed food products, giving a boost to the domestic food-processing industry. The Indian food-processing industry has started looking outward to acquire the latest food ingredients and technology.

India, with diverse agro-climatic conditions, has a production advantage in many agricultural commodities, with the potential to cultivate a large range of agricultural raw materials required by the food-processing industry. India is a major producer of spices, spice oils, essential oils, condiments, and fruit pulps. Significant variations in food habits and culinary traditions across the country translate into a competitive advantage for small and medium local players, who are familiar with local food habits and markets.

The gamut of opportunities that need to be explored and developed in India food processing industry include:

- Dehydration and canning, frozen juice/frozen vegetables, fruit juice and concentrates/pulps, ready-to-serve/eat vegetables, mushroom cultivation and processing,
- Meat and poultry products, integrated abattoirs-cum-meat processing units, frozen and processed sea food,
- Milk products—cheese, frozen desserts, yoghurt, whey, protein, casein, and
- Energy foods/energy/health drinks, wine, snack foods.

Recognizing the importance of the food processing industry, the Government of India has introduced a number of policies as incentives for players to enter and develop the industry. A few of these policy initiatives that offer opportunities for growth of food processing are:

- Food processing industry declared a priority area,
- Full repatriation of profits/capital allowed,
- Almost entire sector is de-licensed,
- Automatic approvals for foreign investment up to 100 per cent, except in few cases and also technology transfer,
- Zero duty import on capital goods and raw materials for 100 per cent export-oriented units,
- Agro-based 100 per cent export-oriented units allowed sale up to 50 per cent in domestic tariff area,
- Export earnings are exempted from corporate tax,
- All processed fruits and vegetables products are exempted from Central Excise Duty,
- Government grant given for setting up of common facilities in Agro Food Parks, and
- Full duty exemption on all imports for units in Export Processing Zones.

Constraints

Sub-optimal growth of food processing industry in India can be attributed to the vicious circle of high unit cost which leads to low demand. This low demand further results in low capacity utilization, ultimately resulting in high unit cost. Domestically, affordability is the key issue. Price differential between fresh and processed food in

India is very high relative to the convenience, hygiene and health values of the processed food. Cost of processed food is high due to bottlenecks in quality of farm produce, availability of infrastructure, credit, technology and equipments for processing and packaging, long and fragmented supply chain, lack of scales, etc.

Cost and quality of farm produce suffer because of tiny and fragmented holdings, low technology, inferior agri-inputs, inappropriate variety, uncertain market and risk aversion, inefficient farm management and cost of logistics. Cost of power, cold chain, water, storage, handling, road and transportation, etc. are high and availability is inadequate or inappropriate.

Presently, India does not have a comprehensive cold chain network. Even though it is the second largest producer of food in the world, the facilities required for food storage and refrigerated transportation infrastructure are grossly inadequate. More than 80 per cent of the cold storage capacity in India is owned and maintained by the private sector. Imports of refrigeration and freezing equipment currently account for approximately 8 per cent of the total market, with the US, Japan, and Finland being the dominant suppliers.

In the absence of properly developed infrastructure for the preservation of raw agri-products, competitiveness and profitability have been eroded. As a result, many plants have become uneconomical and non-functional or they are working much below their potential capacities.

This sector does not often use the best technology, machinery and management practices. Product innovation is sporadic and low. India has been practising various methods of food processing such as sun drying, pickling and fermentation. The quest is forever on for newer methods of food processing with least change in sensory qualities. Irradiation is one of the latest methods in food processing. Irradiation of food items involves treating certain types of food with ionizing energy or radiation. It turns out to be cost-effective method for processing of food products. But, India has still to develop its research base for developing technologies that suit the Indian agricultural and market conditions. A wide network of R&D institutions under ICAR, CSIR, ICMR, and Universities exists in India. But, these are oriented towards academic research and their linkages with users like farmers and industry are undeveloped, resulting in lack of R&D of commercial significance and technology flow. These are not equipped to meet global challenges.

Comprehending this need, the Government of India is no doubt undertaking a number of activities and initiatives. For example, setting-up of CRAFT, the Centre for Research & Applied Food Technology, in 2003 is seen as an important step in supporting India's burgeoning food industry. This centre has been set up specifically to cater to the new and increasing technology and innovation needs of the food processing industry. Nevertheless, right from quality processing technologies to the final packaging and storage, a lot remains to be done.

The success of the food processing industry depends on the existence of an efficient and complete food value chain. Through backward linkages like contract farming and forward linkages like organized retail chains, the food processing industry helps integrate each and every component of the food value chain. This integration is very important for a number of reasons. Apart from reduction in wastages, increased employment and quality products, the integration of the food value chain results in cost reductions and savings at each stage of the food chain. Despite the importance of this integration, the Indian experience hitherto has been disappointing. The food value chain in India as it exists today, is severely fragmented. The food value chain comprises five stages—inputs, production, procurement, processing and retailing. Unfortunately, there are various levels of inefficiencies at each of these stages in the Indian scenario.

In the long and fragmented supply chain, there is a whole plethora of intermediaries starting from the farmer right up to the retailer. In the process, there is a mis-match between demand and supply, limited choice to the consumers, unacceptable wastage and hygiene, avoidable cost addition, opportunistic profiteering and all qualities and grades getting mixed up. The APMC Act, though been amended by a few states is still a major irritant in the supply chain integration.

The food processing industry in India suffers on the retail front as well as due to fragmentation in the retail structure. Marketing and distribution, except by big companies, is largely unorganized and fragmented. This fragmentation has resulted in inefficient storage, transportation and distribution, wastages and value loss. Distribution and intermediation increase the end consumer price by 15 to 30 per cent and must therefore, be minimized. Current distribution structure is risk averse and hence, inhibits stocking new products.

Furthermore, taxes on food commodities in India are very high by international standards. Central and state taxes together increase costs to consumer often by 20-30 per cent. UK, Ireland and Malaysia, for example, have virtually zero tax on food items.

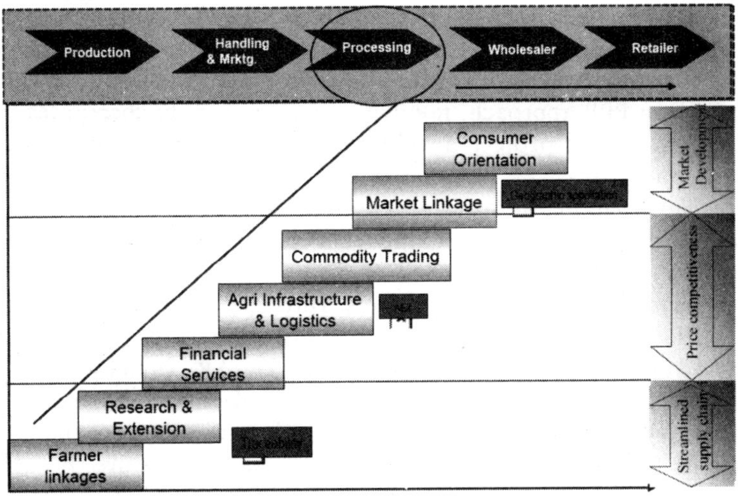

Figure 12.1: Food Value Chain

In India, more than a dozen food laws under several Ministries govern the food sector. These laws are often inconsistent, inflexible and non-responsive to science and innovation. It creates confusion in the minds of consumers, manufacturers and investors and increases the cost of business, affects quality and discourages innovation.

Strategies to Strengthen Food Processing Industry

The time has come to shift the food processing industry from a 'supply' to 'demand' driven sector by reducing costs, enhancing quality and safety systems, building markets, creating and promoting efficient supply chain, developing/inducting world class technology and management and promoting synergy between large and small companies for export and employment. It would require a focused approach to service all stakeholders engaged in the handing of agri-produce, processing, marketing (including exports), infrastructure development, food safety regulation, etc.

Public Private Partnerships (PPPs) must be considered as an important solution to the current weak status of the food processing industry in India. PPPs will help integrate the entire food value chain and ensure that (a) the benefits of the partnership flow down to the rural masses, who have so far been neglected, and (b) the consumers get good quality products at reasonable prices.

The success of the PPP projects in this sector would lie in making the structure bankable. This can be done with formulating projects based on PPP approach, involving more and more corporates in the process of project development and linking them to the consumers.

Furthermore, most farmers do not have access to the demand side as they are not linked to the retail chain or the food processing industry. They cannot expand or diversify themselves as they do not have access to financial markets. Through PPP a special purpose vehicle (SPV) can be created to link farmers in a particular area based on homogenous economic activity. This SPV can be an NGO. The NGO then helps develop projects to link these farmers to the SME segment, where SMEs or large-scale enterprises are the off-takers (any type of food processors). These linkages are through bankable off-take contracts. However, as a majority of NGOs do not have technical and financial skills to carry out extensive feasibility studies and project-related documentation, specialized banks having knowledge of national and international agribusiness should carry out these studies and develop the project framework based in PPP.

Based on the feasibility studies carried out, the banks can then extend loans to the SMEs as well as the micro-sector (SPV), because by linking the farmers with the SMEs and large-scale segments through a project, the bankability increases. The mechanism has been depicted in Figure 12.2.

Developing a viable and bankable project is the key to successful implementation. This requires technical and financial expertise which is not in the skill set of micro-stakeholders or even an SME. Developing the structure for an SPV also involves a special skill set. In this regard, specialized knowledge banks can add value to the stakeholders by providing in-depth understanding of this emerging sector as well as the demand-side linkages with national and international players to actuate the process.

The process of development of the food processing industry would present an array of projects of varying sizes–micro, medium and large–each of which would need to be approached differently. There are two innovative models for maximizing service to the consumer at an affordable price with minimum use of government

The following diagram depicts this mechanism

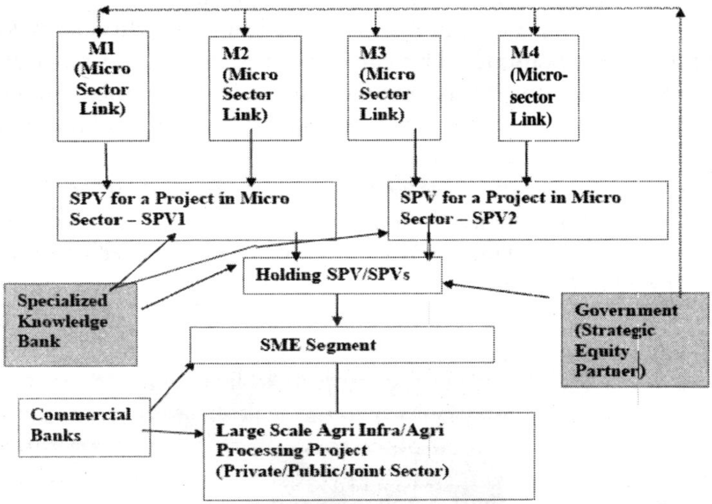

Figure 12.2: Linkages Through Bankable off-Take Contracts

funds, namely the *Soft Funds Model* and the *User Price Rationalisation Models.*

Table 12.1: Soft Funds and User Price Rationalisation Models

Soft Funds Model *Micro & medium scale projects*	*User Price Rationalisation Model* *Medium & larger scale projects*
• Government provides soft funds to the project • Attracts top class private sector participants and private sector efficiencies-knowledge banks such as YBL can attract private sector participants or itself participate in the projects • Project viability enhanced as govt funds do not need to be serviced hence lowering the price borne by the users • Assign/Takeover in cases of deficiency in service	• Typically Build Own Operate (BOO) Model • No upfront investment • Attracts top class PSP and private sector efficiencies • Govt able to use long term planning & resource allocation • Assign/takeover in cases of deficiency in service

Another strategic approach for the development of the food processing industry is the setting up of *hi-tech* farms. A major constraint that is faced by the food processing industry is the availability of good quality and sufficient quantity of farm produce. For example, the mangoes that are easily and readily available may not be appropriate for juices, jams and other processings. Hence, what is required is the setting up of high-tech farms which will carry out extensive and practical R&D for the development of high yield and quality farm produce and also cultivate these products. These hi-tech farms should be undertaken on a PPP basis. A suggestive configuration of a hi-tech farm is illustrated in Figure 12.3.

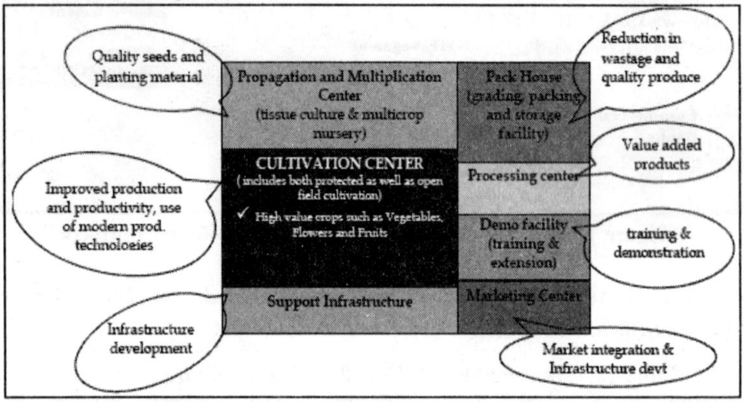

Figure 12.3: Structure of a Hi-tech Farm

Perceiving the industry from a financial institution's (bank) point of view, we have further analyzed how the banking and the food processing industry can create mutually beneficial partnerships. While providing the much needed credit facilities to finance activities along the food value chain, the bank is also benefited through obtaining quality assets to finance the food processing industry as also quality direct agri assets (farm lending).

In the supply chain, there is a huge scope for financing that would help mitigate the usual risks associated with agricultural financing. This would consist of a three-way arrangement—lending to the farmer, arrangement with an off-taker and finally, structuring the deal. A lot of structuring however will be required, which will create retail spin-offs and SME spin-offs. This will create business from end to end. Apart from the regular agri-financing, there will be truck financing or lorry receipt financing, warehouse financing, processor financing and finally when the processed food reaches the wholesaler or retailer, the bank can finance that end as well. Then it gets converted to cash or cheque. In this manner, transactional products and credit products can be created all along the supply chain.

Another very important outcome of the food processing industry is that by assessing the concerned corporate, the bank can actually lend to the farmer(s) who are linked to that corporate. For example, a fruit juice plant owned by a medium or large scale corporate buying quality fruit from a farmer/farmers associations. Here, given that the product (juice) is very well accepted in the market, the bank will not have any inhibitions in financing the farmer/farmer association and will have a quality direct agri asset in its lending portfolio.

A strong case is also present for the banks to lend to ancillary and risky sectors as well. For instance, even though the honey sector is normally considered a risky sector from a bank's point of view, on the basis of an intricate Ware House Receipt Based Structured Trade Finance Solution, YES BANK has recently given loans to over 2000 nomadic honey bee farmers in J&K. The facility provided the farmers with immediate access to credit facilities at reasonable interest rates. The Bank decided to use trade finance solutions in order to overcome the credit risk associated with doing business with small farmers. It created a structure whereby credit risk was transferred from the actual farmers themselves to the stocks of honey kept in warehouses. Once the 'Food Processing Industry' is established, the banking and financial institutions will automatically innovate and develop similar trade finance solutions that will assist the farmer community.

In addition to adoption of innovative PPP approaches, for India to become a major value-added food producer and supplier thereby generating income, employment and foreign exchange, it is imperative for the following sectors to be developed and improved upon.

Raw Materials: Increase in the availability of appropriate variety of raw materials at reasonable price through increased productivity, efficient extension services (both public and private), focused R&D, enhanced credit availability and crop insurance to farmers. Contract farming should be promoted to deal with tiny and fragmented holdings, erratic production due to natural factors, subsistence farming, non-uniform and inconsistent supply of raw materials and long chain of intermediaries.

Infrastructure: Efforts should be made to promote modern integrated cold chains, set up pre-processing centres and pre-cooling facilities near farms and mandis, encourage investor-friendly food parks, upgrade food processing clusters and modernize agriculture markets and warehouses.

Credit: Banks and other financial institutions should take a holistic view of the needs of the entire food supply chain and its unique characteristics. They should ensure adequate and timely credit to investors particularly for the working capital. The government should encourage investment—both domestic and foreign—through an incentive mechanism, simplified procedures and an overall conducive investment climate.

Processing: Existing food processing units and prospective investors need to be encouraged in a big way, to adopt the most efficient technology and processes and best management practices so as to reduce cost, enhance quality and increase marketability. Duty on import of technology and machinery should be reduced, world class technology and machinery should be developed through a network of R&D institutions having a Central apex institute with satellite institutions located strategically in various regions, highly skilled workforce, entrepreneurs and managers should be developed to meet the global competition, agro-processing facilities close to areas of production should be encouraged to avoid wastage and to reduce transportation cost.

Marketing and Distribution: The government needs to encourage modern retailing to eliminate intermediaries and provide large choice and hygienic food to consumers and expand processed food market.

Taxes: Central/State duties, taxes and levies on processed food, machinery, etc. need to be rationalized so as to reach international bench marks and make Indian food commodities globally competitive. Healthy and nutritious food in general should attract near zero taxes. Other food commodities could attract higher taxes depending on their impact on public health.

Market Creation and Intelligence: The state along with NGOs should undertake campaigns for promoting consumption of processed food based on the standards of safety, nutrition and convenience.

Exports: The government needs to undertake a major export drive so as to create export opportunities for farmers and processors where price realization is high and demands are huge. World class technologists and managers must be developed to deal with global challenges and opportunities. Up gradation of ports, airports, storage and handling are essential. Food processing infrastructure in AEZs must be strengthened with a focus on the identified produce. Export promotion bodies must be established for major produces. The government needs to take adequate steps to enable direct farmer–exporter linkage.

Small Scale and Unorganised Sector: For the success of the food processing industry, it is imperative that the small scale and unorganized sector be developed. To achieve this, special arrangements need to be made to provide these sectors with access to efficient technology, knowledge of market and soft credit. Efficient technology for small scale, tiny and unorganised sector must be developed. SSI clusters, anchor industrial centres/or linkages with anchor industrial units to have a network of small and tiny processing units would help the industry immensely.

Conclusions

Food processing is a route to creating sustainable livelihoods and economic development for rural communities. The impact of food processing on all the stakeholders and economy is unequivocally important. Each entity at every stage of the food value chain stands to gain immensely by the development and growth of the food processing industry.

A major impact of the food processing industry is the additional employment that it generates. It creates 1.8 jobs directly and 6.4 jobs indirectly across the supply chain for every Rs 1 million invested. Right from R&D institutes to the farmers to the transporters to the storage companies to the processors and then the retailers, there is additional employment that is being created.

Given India's agro potential, the food processing industry should target to contribute 5 per cent to the world food processing from its current level of mere 1 per cent. For this all the stakeholders including the government and the private players need to actively participate in the development of the industry.

Government policies in the recent past have undoubtedly acted as an incentive for further developments in the sector. There has been a major increase in the amount of investment, both domestic and foreign, flowing into the food processing industry. However, lack of forward and backward linkages in the food value chain, inadequate infrastructure (rural roads, storage, transportation and logistics), inefficient marketing systems and processing infrastructure have acted as a dampener to this industry. These constraining factors require priority attention for harnessing the potential of food processing industry in India, particularly the banking sector must comprehend the potential of the food processing industry and participate in financing strengthening activities at all the stages of the supply chain.

In light of the manifold advantages of this industry for the agrarian-based economy like India, it is the corporate's social responsibility to ensure that the expansion of organized food processing sector benefits not only the private processors and retail companies but farmers also who are the primary beneficiaries and who must be assured of social security and equity.

13

Revamped Rural Credit Delivery for Agrarian Distress Management

V. Puhazhendhi

Agrarian distress on account of deceleration of agricultural growth, since late-1990s, has been recognised as one of the major impediment in the development process of India. The adverse impact of such slow down process is more serious in the rainfed regions that too, among the small and marginal farmers with limited resources. Risk arising out of diversification to high-value commercial crops has further aggravated the agriculture crisis even though a section of big farmers could increase the yields and income through diversification. Diversification has also necessitated borrowing of large amounts of money, increasingly from non-institutional sources at exorbitant rates of interest.

Recognizing the need of increased credit for agricultural development, Government of India (GoI) has initiated series of policy measures. As a result, the institutional credit structure in India has shown a significant growth in both volume and complexity during the past few decades. There is an extensive banking infrastructure comprising 33,411 rural and semi-urban branches of commercial banks over 14,501 branches of RRBs, around 12,000 branches of DCCBs and nearly 1,00,000 cooperatives credit societies at the village level. There is at least one retail credit outlet on an average for about 5,000 rural people, which translates into one outlet for every 1,000 households. This is a remarkable and extensive network. Given this network the moot question would be "Are the financial services needs of the rural poor and vulnerable section comprehensively met by this network and the available financial products are suitable for the existing agrarian distress?"

Recent studies on agrarian distress have revealed that indebtedness is one of the factors linked with farmers' suicides on account of crop failure and related issues. This co-exists with other risk factors like decline in economic status, dent in social position and the inability to meet social obligations. Further, various risk factors can feed into each other and reinforce each other. This situation brings out the fact that the existing institutional arrangement for credit delivery is not adequate and suitable to address the agrarian distress in the country. Thus, the need for revamping rural credit and issues relating to managing the risk through appropriate instruments assumes greater significance in the context of growth process of the country. This Chapter has reviewed the initiatives of different stakeholders for making the institutional credit structure responsive to agrarian distress and has outlined a set of strategies for revamping the institutional credit delivery mechanism.

Extent of Credit Exclusion

The inadequacy of credit outreach by the formal banking system was highlighted by the findings of the 59th round of National Sample Survey Organisation (NSSO) Survey (2003) on 'Indebtedness of Farmer Households'. According to the Survey, of the 89.35 million farmer households forming 60.4 per cent of total rural households, only 43.42 million households (48.6%) availed financial services, especially loans. Thus, 51.4 per cent of the farmer households have either not accessed loans or have been denied loans. In either case, the situation is not tenable given the fact that a large majority of the farmer households possess small farms and cannot be expected to have sufficient funds to finance their farm activities. Similarly, the All India Debt and Investment Survey (NSS 59th Round) has revealed that the share of non-institutional credit in total dues of rural households has increased from 17.5 per cent in 1991 to 29.6 percent in 2002.

Credit exclusion of farm households is the maximum in north-eastern region (80.4%), followed by eastern (60%) and central (58.4%) regions. The problem of credit exclusion is least observed in the southern region (27.3%) where banking infrastructure and banking habits have developed comprehensively. The western region, particularly Maharashtra, which has been facing the problem of farmers' suicides, witnessed credit exclusion by 46.3 per cent of farm households that is below the all-India average. Similarly, in the northern region and group of Union Territories also the credit exclusion was less than the all-India average, at 48.7 per cent and 49.2 per cent, respectively. In terms of absolute numbers, farm households not accessing credit, in the central region draw maximum attention, with 1.58 crore farm households faced with the problem of financial exclusion. This is followed by eastern region having a concentration of 1.26 crore of such farm households. Northern, western and southern regions had credit exclusion to the tune of 53.21 lakh, 47.92 lakh and 44.11 lakh farm households, respectively. The number of farm households facing credit exclusion was 28.01 lakh in northeastern region and minimum level of 36,000 in all Union Territories. An attempt to quantifying the extent of financial exclusion based on the concept of credit desired but denied from the institutional sources, revealed that only 56.8 per cent of the farmer households were able to avail the credit from instructional sources leaving a gap of 43.2 per cent (from non-institutional sources.)

Tribal population is the more severely excluded group, as the incidence of credit exclusion was 63.9 per cent among scheduled tribes. The structure of excluded farm household population in other social groups was: scheduled castes (50%), other backward classes (48.7%) and others (50.7%). Therefore, with the exception of scheduled tribes, all caste groups have exhibited more or less the same average level of credit exclusion. In terms of absolute numbers, farm households affected by credit exclusion in various caste groups are maximum in other backward classes (1.81 crore), followed by others/general caste (1.25 crore), scheduled caste (0.78 crore) and scheduled tribe (0.76 crore).

The problem of exclusion manifests to some extent among all size class of landholdings. However, its incidence was greater among marginal farmer households (55%) than other size categories of farm households, viz. small (49%), semi-medium (41.8%), medium (34.9%) and large (33.6%). The incidence of credit is inversely proportional to the size class of farm holdings. Marginal farmers accounted for a maximum share of 70.6 per cent among the financially excluded as compared to other size categories, viz. small (17.1%), semi-medium (8.5%), medium (3.2%) and large (0.6%) farmer households. The incidence of credit exclusion among farm households was evenly distributed across different economic activities/ occupations.

Incidence of credit exclusion among non-cultivator households (agricultural labourers, artisans, and others) was estimated as 78.2 per cent equal to 4.66 crore in absolute numbers. The number of non-cultivator households affected by credit exclusion was the highest for 'others' category (2.44 crore), followed by agricultural labour households (1.67 crore) and artisans (0.55 crore).

Factors Influencing Credit Exclusion

The factors influencing credit exclusion of farmer households and non-cultivator households including artisans, agricultural labourers and other vulnerable sections in rural areas may be broadly analysed in respect of supply and demand perspectives.

Supply Perspectives: Concern for financial viability of branch is one of the barriers to attracting scattered borrowers for initial access to credit, particularly in the post-reforms period. The policy of branch expansion to rural areas was abandoned in the mid-1990s and the banks were allowed to convert their non-viable rural branches into satellite offices and to close zonal bank branches. RRBs were allowed to relocate their loss-making branches to new places even outside rural areas. As a result, rural branches have steadily come down from 32,981 in March 1996 to 31,967 by March 2005. During the same period, a sharp decline has taken place in the absolute number of agricultural loan accounts. Their numbers declined from 277.4 lakh in March 1992 to 198.4 lakh in March 2001, but started rising thereafter. The trend in the share of agriculture in total outstanding bank credit is similar.

The loan portfolio of commercial banks is tending to focus on short-term credit by avoiding long-term investment projects financing. The increased transaction costs of lending and low net margin have contributed to the perceived weak viability of rural branches, particularly for financing small and marginal farmers.

Poor monitoring of individual borrowers on account of lack of adequate/exclusive staff at branch level is one of the major factors contributing to larger financial exclusion among small and marginal farmers. The attitude of a rural branch is constrained by lack of exclusive staff, sensitization and resources for post-lending processes, which in turn, have a negative influence on new or fresh coverage of prospective borrowers.

Relatively poor recovery performance, especially in the case of government sponsored programmes significantly contributed for the increased financial exclusion among banks. Although net NPAs of banks are coming down, partly due to provisioning and partly due to measures like One Time Settlement (OTS) Scheme, covering past bad loans, banks have tended to become more cautious in lending further to old clients or even new prospective borrowers who may be interested in taking loan for investment.

Demand Perspectives: Indian farming system is still characterized by subsistence farming and about 80 per cent of farmer households do not produce solely for the market. Hence, any efforts on commercialization of small farming households would improve the credit absorption capacity of farmers which in turn increase the demand for credit as well as the existing financial products from the institutional sources.

The problem of credit exclusion has a relationship with infrastructure, which varies across states and regions. The regions having higher incidence of credit exclusion have a lower infrastructure development index. On the other hand, regions having lower rate of prevalence of financial exclusion are the ones, which have higher infrastructure development index. North-eastern, eastern and central regions have relatively lower infrastructure development index, viz., 86.37 per cent, 94.46 per cent (rank 2) and 99.35 per cent (rank 3), respectively, and the incidence of credit exclusion is very high in these respective regions. Southern states with a higher ranking in infrastructure development exhibit the lowest incidence of credit exclusion.

Adequate and effective agricultural marketing structure is another important factor influencing the attitude of non-borrowers in having access to bank credit. Further, a deficiency of establishing a network of marketing at the village level has hindered the expansion of credit. Village level primary markets are not linked with regulated markets. APMC Act has put more restrictions on movement of agricultural produces that facilitated better price for farmers. However, adequate number of regulated market yards and sub-market yards with appropriate infrastructure are yet to be created in all states. Therefore, role of banks, in providing market information to the prospective farmers who would be the clients of banks, is rather limited. Naturally, this creates highly conducive conditions for the evolution of traders, businessmen, commission agents and input dealers in the private markets for widening their network of moneylending activities, which in turn hampers potential growth of institutional credit.

Recognizing the need for greater credit inclusion, especially among small and marginal farmers and vulnerable section of the population, a series of policy initiatives have been undertaken by different stakeholders during the past two decades. Any attempt to review the efforts undertaken would facilitate in outlining the revamping process of institutional credit delivery and emerging alternatives available for future considerations.

Recent Initiatives for Revamping Rural Credit

Recognizing the critical role played by credit in accelerating the growth of agriculture, the Government introduced a policy of doubling agricultural credit in three years during 2004-05. As a result, during the past two years, the number of accounts increased by 36.4 per cent, while the amount increased by 79.4 per cent. This increase in share has been experienced in both, direct and indirect credits for agriculture. However, indirect credit has increased at a faster rate than direct credit. Another important change that has occurred in the recent period is in the size composition of indirect credit. The share of accounts with credit limits of Rs 25 crore and above has increased from 37 per cent at the end of March 2004 to 53 per cent at the end of March 2006.

Initiatives of Reserve Bank of India: The Reserve Bank of India in its Annual Policy Statement of April 2005, has recognised the problem of credit exclusion and, since then has initiated several policies aimed at promoting financial inclusion, especially of pensioners, self-employed and those employed in the unorganised sector. Some of the major initiatives aimed at promotion of financial inclusion include:

- Introduction of a basic 'no frills' accounts. Banking account either with 'nil' or very low minimum balances as well as charges that would make such accounts accessible to vast sections of population. Several banks have already come out with their schemes.
- A simplified general purpose credit card (GCC) to be issued by banks without insistence on collateral or purpose, with a revolving credit limit up to Rs 25,000 based on cash flow of the household to enable hassle-free access to credit to rural households.

- Introduction of a Pilot Project for 100 per cent Financial Inclusion in the Union Territory of Pondicherry and one district each in all States/UTs. On the basis of the experience gained, the scope for providing 100 per cent financial inclusion in other areas/districts would be considered.
- In order to improve provision of financial services in the North-Eastern region and prepare an appropriate state-specific monitorable action plan, RBI has set up a Committee headed by Deputy Governor of Reserve Bank of India, on Financial Sector Plan for the North-Eastern Region with members from financial institutions and banks, state governments from the North-Eastern states and academics.

Initiatives of NABARD

NABARD has experimented several initiatives that have significantly contributed not only to the credit inclusion of poor but also improved their living conditions, especially poor women.

SHG-bank Linkage Programme: Taking into account the apparent inadequacies of the formal financial system to cater to the needs of a significant proportion of the rural poor, NABARD undertook a pilot project in February 1992 to test the potential of homogeneous Self-Help Groups (SHGs) to serve as informal financial intermediaries for improving the access of the poor to financial services. The introduction of the pilot phase of the 'SHG-Bank Linkage Programme' could be considered as a landmark development in the banking for the poor. This owed its origins to the midwifery, if not parenthood of NABARD. As the initial results of SHG-Bank Linkage Programme were promising, the idea gained acceptance in the banking system. NABARD was able to catalyse the banking system in the country to partner informal delivery channels and give the SHG-Bank Linkage Programme a remarkable momentum.

As of March 2006, 2.23 million SHGs comprising 32.88 million poor households were accessing credit through 43,455 branches of commercial and cooperative banks under the SHG-Bank Linkage Programme. Today, the programme is growing at around 6 lakh SHGs annually. During the financial year 2006-07, banks lent Rs 4447.22 crore to 6,12,519 new SHGs and 3,41,598 existing SHGs. The programme is no longer confined to southern states. The non-southern states now account for 46 per cent of 2.23 million credit linked SHGs in the country. Thus, the SHG movement is now a national movement.

At present, there are three main promoters of SHGs: (i) state governments, public sector banks, cooperative banks, (ii) private banks in collaboration with NGOs, and iii) micro finance institutions (MFIs) promoted by non-banking financial orgnistions (NBFCs). There are some instances of the SHGs formed through financial intermediaries (NBFCs) where exorbitant interest rates were charged. Federation of SHGs has been emerging as one of the efficient financial intermediaries. Andhra Pradesh model of Federation of SHGs promoted by the Society for Eradication of Rural Poverty (SERP) provides an ideal model for farmers. Homogeneous group of farmers—size-class-caste-community—could form into small SHGs of 10 to 15 and several such groups formed at a village could form into a federation of SHGs. These village level federations could form into taluka/block level organisations of farmers, which in turn could come together at the next tier as a district level organisation. While the village-level SHGs would access bank credit directly from the banks, higher order activities like, input purchase, marketing, crop/wealth/health insurance could be handled by the block level federations of farmers. The central and state governments may initiate the formation of such SHG federations of farmers and help in their capacity building.

Rythu Mitra Groups (RMGs) in Andhra Pradesh: NABARD in partnership with the Andhra Pradesh State Government facilitated a pilot project for promotion and financing of RMGs during 2004-05. RMGs bring out holistic development in the lives of small, marginal, and tenant farmers through collective action. RMGs besides accessing loans from banks, are expected to serve as a conduit for technology transfer, facilitate access to market information and market, assist in carrying out activities like soil testing, training, health camps, assess input requirements, etc., for its members. During 2005-06, 4437 RMGs were financed by 18 commercial banks, 9 RRBs and 9 DCCBs involving ground level credit flow of Rs 28.11 crore. About 62,000 farmers have been assisted under the pilot project.

Micro Finance Institutions: Following the success of SHG-Bank Linkage Programme and also Bangladesh Gramin Bank model, many of the NGOs engaged in social intervention have taken to financial intermediation for providing financial services to their target clients adopting innovative delivery approaches. Following the RBI guidelines, commercial banks including Regional Rural Banks have been providing funds to MFIs for lending to poor clients. Though initially only a handful of NGOs were into financial intermediation using a variety of delivery methods, their numbers have increased considerably today. While there is no published data on private MFIs operating in the country, the number of MFIs is estimated to be around 800. A large majority of MFIs operate on much smaller scales with clients ranging between 500 and 1,500 per MFI. However, a few big NBFC MFIs have an outreach of over 100,000 clients. A growing number of NGO-MFIs are keen to transform into NBFCs in order to offer financial services to the poor, while others are forging agency agreements with diverse partners including insurance companies to offer important risk management services.

With the objective of ensuring greater credit inclusion and increasing the outreach of the banking sector, banks have been allowed to use the services of Non-Governmental Organisations (NGOs), Self-Help Groups (SHGs), Micro Finance Institutions (MFIs) and other Civil Society Organisations (CSOs) as intermediaries in providing financial and banking services through the use of Business Facilitator and Correspondent models and provisions for correspondent banking have opened new and diverse avenues to address the issue of financial inclusion by banks.

Joint Liability Group Approach: Innovations like SHG-Bank Linkage Programme has proved to be successful in providing financial services to assetless or ultra poor. In order to develop effective credit products for the mid-segment clients having access to productive assets, NABARD had piloted the project during 2004-05 in 8 states of the country through 13 RRBs through the mechanism of joint liability approach. A Joint Liability Group (JLG) is an informal group comprising 4 to 10 individuals coming together for the purposes of availing bank loan either singly or through the group mechanism against a mutual guarantee. The JLG members offer an undertaking to the bank that enables them to avail loans. Unlike SHGs, which are more process oriented, the management of the JLG is simple with little or no financial administration within the group. The JLG experiment with select banks has demonstrated that banking system can reach clients like tenant farmers, share croppers, oral lessees, farmers with small holdings without proper land records through the joint liability approach. As on March 2006, the participating banks promoted 870 JLGs and extended finance aggregating Rs 10.88 crores to 494 JLGs. Based on the successful experience, NABARD has finalized the guidelines for mainstreaming the programme with commercial banks on an all-India basis.

Watershed Development Approach: In India, nearly two-thirds of 142 million hectares of land is under rainfed cultivation, which contributes about 45 per cent to the total food grain production. As against this, the irrigated areas account for only one-third of the cultivated area and contribute 55 per cent to total food grain production. The productivity of land in rainfed areas is quiet low, at around one-third of the irrigated areas. Low income is one of the reasons for financial exclusion. NABARD has therefore, adopted the watershed development approach with community participation for increasing production and productivity of crops and thereby, leading to higher employment and income generation opportunities.

NABARD is involved in two major programmes on watershed development since December 1992 under the Indo-German Watershed Development Programme. Under this programme, 444 villages in Maharashtra and 45 villages in Andhra Pradesh involving 32,300 farm families were covered. The programme adopted a participatory approach wherein the community residing in the area had a major role in planning, implementation, supervision and monitoring of the programme besides maintenance of the common assets created under the project. The community was also given the responsibility of managing the funds, resolving conflicts and taking all project related decisions. The services of experienced NGOs were provided to the villagers to guide them in project implementation as also for imparting training from time to time and providing technical guidance.

The participatory watershed development approach showed excellent sustainable results in nearly 100 watershed projects. As a result, NABARD considered it necessary to replicate the message of participatory watershed development to other states for the benefit of poor people residing in rainfed areas across the country. For this purpose, Watershed Development Fund with a corpus of Rs 200 crore contributed equally by GOI and NABARD has been set up.

The Fund aims at fostering implementation abilities at the local level and creating community infrastructure for micro watershed projects through the active involvement of Gram Panchayat, community based organizations like, NGOs, SHGs, to spread the message of participatory watershed development throughout the country. So far, 427 projects have been financed in 12 states covering 2624 villages and benefiting 1,28,100 farm families involving 4.03 lakh hectares. A total disbursement to the tune of Rs 2940 lakh has been made up to 31 March 2006. This includes grant to the tune of Rs 2440 lakh and loan component to the state governments to the tune of Rs 500 lakh.

Many of the state governments have included watershed activities as one of the eligible item of works under National Rural Employment Guarantee Programme. The initiative of state governments for convergence of watershed activities and NREGP has also been a step in financial inclusion as wages are credited to the savings account of labourers.

Tribal Development Programme through Wadi Approach: NABARD has been closely associated with the implementation of an Adivasi Development Programme through 'Wadi' approach in collaboration with NGOs in Gujarat since 1995 and in Maharashtra since 2000. The central focus of the programme is on "wadi"—which means small orchard—with other development interventions like, soil conservation, water resource development, community health, sanitation and women development as supporting components. A special feature of the programme is the blending of grant with credit not only to ensure participants' stake and involvement in the programme, but also for self-reliance of the participants.

In these two states, over 27,500 tribal families have been covered under the programme and an area of around 24,000 acre, spread over 400 villages has been brought under 'wadis' of cashew, mango and amla alongwith boundary plantation of fuel wood and fodder species. Production has already started in the Gujarat programme with around 300 Mt of cashew and 500 Mt of mango, which has been instrumental in establishing 10 village level processing units providing perennial employment to 180 landless tribal people, most of which are women. Ten tribal co-operative societies have also been registered to facilitate the continuation of the programme once the external support is withdrawn. A shift in cropping pattern towards commercial crops has resulted in generating regular income. This has helped in developing saving habits through SHGs and has led to reduced dependence on moneylenders. The wadi model merits wider replication for poverty alleviation as it is targeted at *advisies* who are most vulnerable for financial exclusion.

Kisan Credit Card: The introduction of the Kisan Credit Card (KCC) in 1998-99, a government initiated measure, was a step intended to overcome the rigidities inherent in the credit system and to make the credit market more borrower-friendly. The scheme aims to provide adequate, timely credit support to farmers from the banking system in a flexible, hassle free and cost effective manner. The farmers may use the card for the purchase of agricultural inputs and other production needs. Credit limits are fixed on the basis of operational holding size, cropping pattern and scale of finance. The entire credit needs for a full year including ancillary activities related to crop production, such as the maintenance of agricultural machinery/implement, electricity charges are covered. All Public Sector Commercial Banks, State Cooperative Banks/District Central Cooperative Banks and RRBs are implementing the KCC Scheme in all states and union territories. By the end of 2006, the number of cards issued has been reported to rise to 642 lakhs.

Review of Relief Package for Agrarian Distress

The Prime Minister announced a comprehensive package to provide relief to the distressed farmers in July 2006. It addresses the causes for distress and provides scope for rehabilitation and relief to farmers in the 31 identified districts spread across the four states of Andhra Pradesh, Karnataka, Kerala and Maharashtra.

The rehabilitation package for the four states is to the tune of Rs 16978.69 crore consisting of Rs 10579.43 crore as subsidy/grants and Rs 6399.26 crore as loan. The package is to be implemented over a period of three years, from 2006-07 to 2008-09. It includes both, short-term and medium-term measures. The credit component includes improved supply of institutional credit, waiving of the entire interest on overdue loans as on 01 July, 2006 and provision of fresh credit by the banking system, increase in supply of credit and rescheduling of overdue loans over a period of three-to-five years, with a one-year moratorium. The central and state governments will share the burden of waived-off overdue interest equally. The non-credit component of the package aims to revive the livelihood base of the distressed farmers. Schemes involving irrigation, watershed development, diversifying agriculture towards horticulture and linking it with agro and fruit processing, development of allied and non-farm sector activities are to be implemented.

The implementation of the package provides for state-level committees consisting of central and state governments representatives concerned with coordination and supervision, district-level committees and Panchayati Raj Institutions (PRIs) at the local level for delivery of the package and optimum utilization of resources in a time-bound manner. The credit components of the rehabilitation package are being implemented by the Commercial Banks, Regional Rural Banks (RRBs), and Cooperative Banks.The Ministry of Finance, Government of India, is supervising and monitoring its implementation.

The credit component of relief package was a short-term measure to give relief to the farmers immediately. The rationale behind the credit component was to build an efficient and viable farming system by providing adequate credit at low cost. The credit requirement for the farmers in these districts was assessed based on the available potential and credit absorption capacity of farmers as indicated by the Potential Linked Plan (PLP) prepared by NABARD.

The achievements under the credit component are enumerated below.

- Against an overall target of Rs 2718.25 crore for the four states, the achievement is Rs 3535.19 crore (130%) under waiver of interest on overdue loans as on 01 July, 2006. Andhra Pradesh, Karnataka and Maharashtra have overachieved their targets. Kerala achieved 90 per cent of its target.
- Principals to the tune of Rs 10489.50 crore have been rescheduled against a target of Rs 9140.19 crore and against this rescheduled loans,
- farmers have been given fresh loans of Rs 16592.83 crore for these states taken together. In the case of reschedulement of loans, all the states barring Kerala have achieved their respective targets. Kerala achieved only 42.55 per cent of the target allocated to the state.
- Fresh loans to the tune of Rs 16592.83 crore have been disbursed against a target of Rs 21422.44 crore for these states taken together. Interestingly, Kerala achieved cent percent target in disbursing fresh loans whereas the other three states achieved two-thirds of their targets.

The non-fulfillment of target under fresh credit in the remaining three states could be attributed to migration of farmers, selling of their land or giving up farming altogether. The gap in the off-take of fresh credit in all the three states (AP, Maharahtra and Karnataka) points to the fact that the credit needs of the farmers perhaps were not assessed accurately at ground level. In the Vidharbha region, the credit flow targets fixed should have been realistic based on the proper assessment of the credit absorption capacity at farm level/household level. The doubling of targets fixed for credit disbursement in the region is likely to increase debt levels and adversely affect the credit flow next year. In order to ensure that the basic objectives of providing farm credit are not distorted, disbursements should have been made only after proper appraisal of the proposals. This also calls for greater co-ordination among banks, block level officials at the ground level in identifying the genuine credit needs of the people.

Issues in Implementation of Relief Packages

Broadly the relief package has attempted to put together a comprehensive plan in place. Though the design part of the relief package has been conceptually well-drawn, uniform application in different states without recognising the location-specific situations has led to certain implementation problems. In other words, top down approach has been followed while implementing the package. For example, the interventions like Watershed Development, Check Dams, etc. proposed for the districts in Kerala may not be very effective as distress in these areas has arisen more out of the price fluctuations. Not all the districts identified for the package are rain-dependent, for example, all the identified districts in Kerala and districts of Shimoga and Kodagu in Karnataka receive rainfall above 2100 mm. Moreover, the crops grown in these districts also vary widely with cotton and soyabean occupying major areas in Vidharbha, groundnut and rice in Andhra Pradesh and Kerala growing plantation crops in all the three districts. Moreover, the distress is not confined to small and marginal farmers across the districts, in the case of Shimoga and other plantation growing states of Kerala, it is the problem of big farmers whose expectations and life-style are poles apart from those of the small and marginal farmers. Thus, the strategy to mitigate distress needs to be region-specific and not uniform across the districts.

Though the package attempts to synchronize credit, productive capacity of land-based assets and supplementary income generation activities, the sequencing of different parts of the package needs improvement. Any attempt to target the credit flow without adequate work on irrigation, backward and forward linkages would not achieve the desired results. Therefore, there needs to be a clear consensus among various stakeholders that more planning should have been done before introduction of the package. This mid-term correction can be undertaken by involving actively the ground level functionaries and the community.

A number of departments are associated in the implementation of various schemes under the project. Therefore, co-ordination among all the central and state departments needs to be strong to increase the pace of the project. Lack of monitoring and proper management of information system for both, the credit and non-credit components of the relief packages, were one of the hindrances in assessing the progress and impact of the packages at the grass-root level. Lack of follow-up by different stakeholders while implementing the relief packages affected the benefit to be recieved by the target clients.

Future Strategies

The problem of financial exclusion is acute, as about 51.4 per cent farmer households do not have access to or do not avail credit from any source–institutional or non-institutional, formal or informal. This phenomenon is influenced by both, credit and non-credit related factors which include: primary concern for branch viability, increased cost of transaction, poor monitoring of individual borrowers, lack of exclusive staff in branches, cautious attitude, low suitability of loan products, lack of non-financial services, small subsistence farming, and lack of infrastructure facilities. A revitalized consideration to these factors would help in giving a boost to financial inclusion through credit.

Integration of Domestic and Global Markets: In the long-run, rejuvenation of the Indian agriculture lies in addressing basic structural, institutional and technological factors as much as restructuring state support systems in the face of growing exposure to local and global market forces. In the context of rapid modernization of agricultural holdings, the focus of strategy for revival of agriculture should be on the small-farmer economy.

Diversification in Agriculture: The declining trend in public investment should be reversed, and public investment in research and development with adequate priority to rainfed agriculture should receive due priority. Public institutions, like seed corporations should assume greater role in making available critical inputs, such as seeds and extension services. Finally, there is no way that small and marginal farmer households can improve their living standards by depending solely on the agricultural income. There is a need for diversification into off-farm and non-farm activities, which should account for their household income. Rural farm–non-farm linkages still remains at a very low level, and require appropriate policy support to gain momentum. There is a need for optimum combination of both, farm and non-farm activities and coordination of efforts in terms of training, capacity building, asset creation and market improvements.

Institutional Arrangements: There is a need for strong grassroot level institutional arrangement at the village level for the effective implementation of the relief packages. The SHGs and Farmers Clubs are gaining greater acceptance at the village level; such efforts need to be followed covering all the farmers in the villages. SHG movement is very predominant in Andhra Pradesh, their successful experiment such as the Rythu Mitra Groups, etc. may be appropriately replicated and adopted in other regions. Micro Finance Institutions will have a greater role for increased outreach of credit to rural poor and can serve as effective intermediaries between banks and financially excluded farmers.

Innovative Financial Products: The validity and suitability of the existing financial products offered by the financial institutions need to be reassessed based on demand. The products, which are provided by the financing institutions, must be made more attractive to the vulnerable groups. There is a need for integration of financial and non-financial services with its optimum mix to increase the demand of the products given by the financial institutions. While designing the products, the financial institutions need to re-define its approach to cater to the needs of small and marginal farmers. Strategies need to be developed for marketing the existing financial products and making them accessible to the vulnerable groups.

In the short-run, concrete measures have to be undertaken to make available timely and adequate institutional credit at reasonable rates of interest for undertaking productive expenditure including basic consumption needs in the lean period, and to reduce the debt burden of vulnerable sections of the population.

Adequate crop insurance requires very high premium making it unaffordable and any subsidy to premium increases burden on the state-exchequer. It means that the only way to administer credit is to push it through buy back arrangements, contract farming, organised marketing, processing, retailing or exports in a project mode with bankers, efficient marketing chain, processors and exporters wherein returns to the farmers are also proposed to be increased to enable them to service the loans through technology dissemination, extension, diversification and value addition.

Revival of Cooperatives: Several committees have gone into the question of reorganizing the cooperative credit structure in the country. Recently, the Task Force on Restructuring of Cooperative Credit Institutions (2005) underlined the need to eliminate state interference in the functioning of cooperatives. It has recommended a revival package of about Rs 15000 crore for retiring the share capital contributed to the co-operative societies by state governments and for cleaning up their balance sheets. The report has also recommended capacity building, human resource development, institutional restructuring to ensure democratic functioning, and improving the regulatory regime to empower RBI to enforce prudent financial management.

The Task Force has also recommended that rural financial cooperatives should be dealt with as a distinct and separate class and should be made fully democratic, self-governing, self-reliant organisations for mutual thrift and credit. It has been suggested that cooperatives be reorganised primarily as credit and marketing societies. Implementation of the Task Force recommendations would go a long way towards reviving cooperative credit societies in India. The Government of India has accepted most of the recommendations of the Task Force and has already put in place a revival package for cooperative credit. There is also need for a revival package for long-term cooperative credit.

References

NABARD, *Annual Report* (Various Issues), National Bank for Agricultural and Rural Development, Mumbai.

NSSO (2005), *Situation Assessment Survey of Farmer Households*, Report No.498 of NSSO, Ministry of Planning and Programme Implementation, Govt. of India, New Delhi.

RBI (2004), *Report on Trend and Progress of Banking in India 2003-04*, Reserve Bank of India, Mumbai.

Section V

POLICY IMPERATIVES

14

Policy Imperatives for Addressing Agrarian Distress

Smita Sirohi and B C Barah

Introduction

Though agrarian distress is a reflection of multiple causes, chronic indebtedness of rural households is considered to be vital manifestation of farmers' distress and precursor of suicides. Examining the issue of agricultural distress for suggesting measures to assist distressed farmers, the Johl Committee (RBI, 2006) has observed:

> *"The manifestation of distress is stressful behaviour arising out of social, economic and psychological reasons. However, one common factor that can be seen across all regions is that manifestation of economic distress is primarily through indebtedness. "*

The overwhelming feeling in the policy circles that large scale defaults in loan repayments are an indication of the abject economic situation of the farmers which has culminated in the Farm Loan Waiver Scheme of Rs 600 billion (subsequently raised to 716.8 billion) as major policy response to address agrarian distress. There are, however, diverse opinions regarding the efficacy, utility and economic rationality of the Scheme from the perspective of lightening the debt burden of the farmers, financial health of the formal credit institutions, and reaching out to the root of the problem of agrarian distress. Hailing the Scheme, the Prime Minister Manmohan Singh described the agricultural loan waiver scheme as a 'very unorthodox response' to raise the depressed 'animal spirits' of the farmers, the 'biggest businessmen' of the country. On the contrary, Mahendra Dev (2008) has argued that the negatives in the form of exclusions and limitations outweigh any positive outcome of the Scheme. The waiver is not considered as a solution to the crisis even in the short-run let alone long-term agrarian problems (Sainath, 2008).

Whatever may be the merits or drawbacks of the loan waiver policy of the Government of India, one thing is for certain, viz. the loan waiver cannot be expected to automatically enhance farmers' capacity to increase production and incomes. The farmers need measures for raising output and good prices for production rather than more credit which, in the absence of viable agriculture, push them back into debt trap. As has been emphasized by several committees, indebtedness is only a symptom and not the cause of widespread agrarian distress, which should be addressed in its totality. In the past few years, several commissions, committees and working groups have given a number of recommendations for the revival of agriculture. This Chapter synthesizes the various policy options for tackling the emergent situation and making agriculture profitable for secured livelihoods and faster economic growth. The remedial measures to alleviate agrarian distress have been discussed under seven broad themes identified as: (1) TFP Enhancement, (2) Infrastructure Development, (3) Risk Management, (4) Provision of Financial Services, (4) Institutional Reforms, (5) Social Support Mechanisms, (6) Non-farm Employment and (7) Human Development.

Enhancing Total Factor Productivity (TFP)

Maintaining the growth of TFP is vital for sustainability of a production system. The positive impact of TFP instead of agricultural income growth on agrarian distress is more discernible as it signifies technology-driven shifts in the production function rather than rise in income from simply increased input-use.

R&D Investment

Fan *et al.* (1999) have shown that R&D investment has the highest marginal impact on TFP of the agriculture sector. Undoubtedly, technological progress in agriculture has spearheaded the growth momentum in the sector, particularly after the mid-1960s, but unfortunately the investment intensity ratio of public Research and Extension (R&E) is lower in India (0.4 per cent) than the average intensity of 0.6 per cent in developing countries (0.6 per cent) and 1 per cent at the global level (Pal and Byerlee, 2003). Increasing the investment in agricultural research is therefore vital for increasing TFP, as the present level of research intensity is inadequate to meet the emerging R&D challenges on the national and international stages. Voices have often been raised to increase the government investment on research to at least 1 per cent of the agricultural GDP (Pal *et al.*, 2000), particularly to rationalize resource allocation across regions and commodities.

Besides being under-funded, agricultural research is also not well targeted. Available resources are not being optimally utilized because of lack of a clearly stated strategy which prioritizes the research agenda rationally. A major paradigm shift is needed to transform the present commodity-based research to a systems approach. Since farm-level problems are specific to agro-climatic zones (ACZs), what is needed is a convergence between R&D agencies within individual ACZs so as to bring region-specificity in technologies and their time-bound assessment.

Input Delivery and Management

In addition to increasing investment in agricultural research and development, timely availability of quality inputs and efficient management of inputs are vital for stimulating growth in TFP. Poor input delivery system and environmental problems associated with irrational input management are near universal factors of low productivity in almost all the agro-climatic regions (Box 1).

Seed: Seed is a trigger point which sets in motion the process of technological change. At present the Seed Replacement Rates (SRRs) in respect of various crops in different states are still relatively low and need to be raised. The desirable SRRs, without which it is not possible to achieve higher productivity, are 25 per cent for self-pollinated crops, 35 per cent for cross-pollinated crops and 100 per cent for hybrids (GOI, 2008). The distribution of certified seeds has increased more than two-fold during TE 1996-97 to TE 2006-07 from 79.01 lakh quintals to 191.98 lakh quintals, but the quality of these seeds is far from satisfactory. Even in case of genetically modified seeds of Bt cotton that are four-times more expensive than the normal hybrid varieties of seeds, germination of only 65 per cent is guaranteed. There are also reports that seed sown led to healthy growth of the plants but did not yield any produce (RBI, 2006). The huge difference in prices between ordinary and technology intensive seeds, increases the risk of spurious seeds filling the middle price range. The state machinery should, therefore, have the wherewithal to deal with deviant behaviour on quality and regulatory framework needs to be put in place for ensuring reasonable seed prices to the farmers.

Irrigation: Irrigation is another important pillar for realization of technological gains. There are a wide range of issues in irrigation that require policy attention such as.

Box 14.1: Region-Specific Factors Causing Low Productivity

Agro-climatic region	*States/Parts of States*	*Region-specific constraints*
Western Himalayan region-I	J&K, HP, Uttaranchal	Severe soil erosion, degradation due to heavy rainfall/floods and deforestation, low SRRs, poor input delivery, inadequate communication infrastructure and marketing
Eastern Himalayan-II	Assam, NE states, Sikkim	Aluminium toxicity and soil acidity, soil erosion and floods, shifting cultivation, low SRRs, nonavailability of electricity, poor road, poor input delivery system and communication infrastructure
Lower and middle gangetic plains regions-III and IV	West Bengal, Bihar, Eastern UP	Flood/water logging, improper drainage, salinity/alkalinity, arsenic contamination, low SRRs, nonavailability of electricity, high population growth, poor road and and communication infrastructure
Upper and trans-gangetic plains region-V and VI	Western UP, Punjab, Haryana	Groundwater depletion, decreasing total factor productivity, micro-nutrient deficiency, nonavailability of electricity, and high population density
Eastern plateau and hills region-VII	Orissa, Jharkhand Chhattisgarh	Moistur stress, drought, and soil acidity, iron toxicity, low SRRs, nonavailability of electricity, high population growth, poor road, poor input delivery and communication infrastructure

Source: Cited in Report of the Working Group of Sub-Committee of National Development Council on Agricultural and Related Issues on Region/ Crop Specific Productivity Analysis and Agro-Climatic Zones, Planning Commission, Government of India (February 2007).

- Increasing investment for creation and utilization of irrigation potential in states of Andhra, Gujarat, Bihar, Tamil Nadu, Karnataka, Maharashtra, Madhya Pradesh and Orissa that have high magnitude of untapped potential,
- Improving water-use efficiency through renovation and moderization of canals that would reduce conveyance and seepage losses
- Governance to prevent over exploitation of groundwater resources
- Ameliorative measures to check water logging, salinity and alkalinity in canal irrigated areas
- Rational pricing of water

Agro-chemicals: Judicious use of fertilizer and plant protection chemicals have important role to play in soil and plant health management. The Eleventh Plan document recognizes that irrational system of fertilizer subsidy has become counter—productive as distortions in prices ratio in favour of nitrogenous fertilizer has led to imbalance in the use of nitrogen–phosphorus–potassium (NPK) creating serious problem of soil degradation and adversely affecting productivity. The policy should therefore, focus on restructuring the subsidy to induce the use of NPK in right proportion.

Similarly, for preventing indiscriminate use of plant protection chemicals, policy interventions for promoting precision agriculture would not only bring down the cost of input use but also make them more effective for enhancing crop productivity.

Infrastructure Development

Agricultural output growth is determined largely by technology and prices, in combination with agricultural infrastructure. Government has traditionally been well aware of the critical role of infrastructural facilities in agricultural growth and has hence high priority has been accorded to its development since the beginning of planning in the country. A recent policy initiative to build rural infrastructure is the ambitious Bharat Nirman Yojna, launched by the Government of India in 2005, which aims to provide electricity, all weather roads, drinking water and telephones to all the uncovered villages.

In the changing economic scenario, there is a need for a new approach for infrastructure development, both, in terms of (a) promoting new forms of infrastructure that have hitherto not received prime attention so far, and (b) looking for alternative sources for financing infrastructure, particularly through enhanced private participation. While roads, electrification, irrigation, credit network and other basic infrastructure are still important and continue to be inadequate in their coverage and efficiency; new forms of infrastructure such as information and communication technology, cold chains, model terminal markets, meteorology stations, etc. are emerging and becoming pre-requisite for future sources of agricultural growth and poverty alleviation. Similarly, fiscal stringency, the increasing requirement to provide efficient infrastructure services in a globally competitive set-up and rethinking on the ability of the government owned entities to supply 'quality' infrastructure makes it imperative to attract private entry in provisioning of infrastructure.

Inadequate investment and ineffective governance mechanism are two critical factors responsible for insufficiency of infrastructural availability both, in terms of its physical access and quality and the regional imbalances in the spread of infrastructural facilities. One of the critical areas of concern in infrastructure development is poor quality of services which has its roots in policy overemphasis on construction rather than maintenance of the created facilities. Creation of infrastructure is not followed by adequate institutional and financial mechanism for operation and maintenance of the same. Rural countryside is dotted with potholed roads, decaying bridges, electricity lines that carry no power, dry water supply schemes, crumbling school and hospital buildings. The nature of economic activities and business environment and the expectations and aspirations of the masses is undergoing a sea-change, due to which the issue of the sustainability of the created infrastructure needs visibility in policy space.

In light of limited financial resources and demand for a wide range of infrastructural facilities, the development policy needs to prioritize public investment for realizing the benefits of infrastructure development for the agriculture sector. Based on the available studies on the impact of the different forms of infrastructure and constraints in their development it can be said that reliable power supply and road network facilities lie at the foundation of comprehensive rural infrastructure provisioning. In the rainfed areas, irrigation is to be among the top three infrastructure that have to be developed. Improvement in communication and information technology infrastructure have the potential to change social and economic patterns of rural India as well as provide poorer country—dwellers with information they can use to improve their lives. Once roads and ICT provide connectivity to rural masses, the development of modern marketing facilities for transportation of fresh produce to urban consumers in India and abroad can improve the agriculture sector's productivity and raise standard of living of millions of Indians living in villages. Integrating the agro-sector with the world economy will pull many village folk out of distress and poverty and provide them productive employment.

In the context of promoting private entry in infrastructure sector, with subsequent stages of infrastructure development the government needs to redefine its role vis a vis the private sector and infrastructure users in the following way:

In the next stage of development, the government continues to be the regulator but finances infrastructure to a limited extent. Instead it is the public or private entity that makes use of financial instruments for investing in infrastructure and also carries out the operational responsibility. The investments for infrastructure creation come from the public/private entity but the O&M cost is totally borne by the users who also have a voice and grievance redressal for assured supply and service quality. In the final stage, the government is only the regulator and provider of viability gap financing, whereas users are well-organized into a representative body which finances O & M and operates the infrastructure facility. The private entity is the builder and major financer.

The policy measures that will have to be put into place for enhancing private participation include relaxing the financial regulatory regime for enhancement of financial flows for infrastructure projects, new financial innovations like, securitization for catalyzing infrastructure financing in the country (Singh, 2008), income tax incentives, smart use of subsidies and creation of special purpose vehicles.

In using subsidies for infrastructure, the Governments often face the dilemma of giving subsidies to meet user charges or as financial instruments to attract private investment (Mehta *et al.*, 2003). Thus, in the context of market-oriented economic policies, it would be more rational to give subsidies for developing sustainable markets for the private provision of rural infrastructure. The design of the subsidy disbursed is however, crucial to how effectively infrastructure is built, financed and managed. Outright and upfront grants have often failed to motivate the beneficiaries to use the funds judiciously. Therefore, what is required is the smart use of subsidies such that the grant is targeted appropriately to ensure efficient and sustainable use of funds. The output based aid schemes is one such way of granting subsidies, where the payment of subsidies is linked to the outputs or results actually delivered. Another way in which the earmarked government funds can be employed more creatively for infrastructure financing is their used as equity or quasi-equity, or even as a guarantee.

Risk Management

Reducing exposure to risk and improving household capacity to manage risk, both ex-ante and ex-post, can encourage greater investments in productive activities, selection of technologies with greater expected returns and entry into high value markets thus, being instrumental in relieving the distress situation faced by the farmers. However, lack of effective risk mititgation products is the biggest missing link in agriculture. Some of the ex-ante risk management options where direct instruments shall be applicable are: (a) formal and informal insurance mechanisms (b) futures and forward markets; while the ex-post risk coping mechanisms that are to be dealt with by way of institutional reforms include: (a) safety nets and (b) more efficient, accessible and stable asset and labour markets (Acharya, 2006).

Crop insurance was introduced in the country two decades ago to hedge the farmers against yield risk. In its present version, the National Agricultural Insurance Scheme (NAIS) has been available since 1999–2000 but despite the utility of the Scheme, its coverage is extremely limited. From the users' perspective the Scheme is not popular due to unrealistic crop loss assessment and from the angle of service provider it is not financially viable as premiums are not based on actuarial principles. Premiums for most important crops are fixed at all-India level irrespective of risk and the government pays for entire excess of claims over premium received. The research and policy focus for agricultural insurance should therefore be at developing and implementing insurance products that adequately address the interests of users and suppliers of these products.

Availability of reliable information on the weather parameters is a prerequisite for success of crop insurance. Some private players in the agricultural insurance sector like, ICICI has about 100 automatic weather stations (AWS) installed in MP, AP and parts of Karnataka to collect reliable weather information for its rain insurance cover. But overall network of AWS in the country is very thin as only about 60 per cent of the 3000 odd AWS installed in the country are functional. Moreover, most of AWS are under public organizations and institutions such as, IMD, KVK, SAU, Regional stations, Research Institutions and State Agricultural Departments that hardly have any linkage with the insurance service providers. The policy attention for widening and strengthening the AWS network would be an important catalyst for agriculture risk management. The *Report of the Expert Group on Agricultural Indebtedness* has suggested facilitating the crop surveillance mechanism by making using of remote sensing data and sharing it with concerned agencies associated with relief and risk insurance.

In addition to yield risk, price risk is also very important in agriculture. To hedge against price risk, the Government of India has been providing minimum support prices across 24 major crops. This reduces the seasonal risk faced by farmers to some extent and small and marginal farmers are also somewhat cushioned by the minimum support prices. However, due to increasing integration with the global market, the farmers are now subject to greater price risk because of variability in world prices. This calls for putting in place more effective price risk hedging mechanisms than just the MSP. Commodity futures market is an important step for providing some respite to the farmers. In the last couple of years, the Ministry of Consumer Affairs, Government of India, and the Forward Markets Commission, which is also the immediate regulator, have opened the gates for futures trading in over 120 commodities with the basic objective of presenting to the producers, suppliers and end-users an opportunity to hedge their price risk. But direct participation of farmers is negligible and price volatility does not appear to have reduced. Other than the initial teething troubles in operation of these markets, quality specifications, delivery norms, margin and lot-size of most commodities traded at the bourses make it difficult for the average farmer to directly participate in exchange trading as hedgers. As the trading lots are often much bigger than what is produced by a small farmer, policy intervention is desirable for enabling the banks or other institutions to assume the role of an aggregator and integrate smallholders to commodity futures market.

The formation of price risk mitigation fund is another policy option recommended by the Radhakrishna Committee to mitigate the impact of price collapse in cases of commodities not covered by minimum support prices. Similarly, the M S Swaminathan Commission also provides recommendation for setting up a price stabilisation fund.

Management of Credit

Indebtedness being one of the vital manifestation as well as reason of farmers' distress, policies for prudent financial management have received emphasis in the recommendations of various advisory bodies constituted to examine the issue of agrarian crisis in India. The Expert Panel constituted by NABARD recommended a debt swap scheme to redeem high cost moneylender debt. Drawing from the examples of State Level Banker's Committee (SLBC) and Indira Kranti Patham (IKP) Scheme in A P, the Panel maintains that although debt swap does not reduce the indebtedness of the household it makes the payment terms less onerous.

The Working Group on Distressed Farmers (Johl Committee) suggested introduction of credit guarantee scheme (CGS) for borrowers with small loans/credit limits up to Rs. 1.00 lakh, to be administered by the Deposit Insurance and Credit guarantee Corporation. CGS is cap on the extent of guarantees, which is generally linked to the net owned funds/capital of the fund. There is a system of specific guarantee in place after assessment of credit and risk evaluation. The main features of the proposed CGS are summarized in Box 14.2.

Box 14.2: Features of Proposed Credit Guarantee Scheme

- All the commercial and rural cooperative banks have to compulsorily participate in the Scheme.
- Covers only the borrowers affected by 'systemic distress' (extensive loss of crops/assets caused due to natural calamities and pests/locusts on an epidemic scale).

Contd.

Contd.

- Covers those borrowers with aggregate sanctioned limits up to Rs 1 lakh granted after introduction of the Scheme, and whose loans are required to be restructured/ rescheduled for the second successive time on account of systemic distress.
- Guarantee up to 60 per cent of the amount outstanding in the guaranteed account/s, as on the date of occurrence of the natural calamity, necessitating restructuring/ reschedulement for the second successive time. The balance loss should be borne by the bank/s concerned.
- Credit guarantee available only in respect of the borrowers where the crops/assets have been insured.

Another important aspect in credit management that deserves policy attention is provision of services like, financial counseling to the farmers. Low literacy levels coupled with asymmetry in the access and flow of information often constraints the farmers in efficient management of finances in hand. Therefore, if the agricultural/rural development officers with the State department or banking institutions could also be trained and sensitized to cater to the latent counseling needs of the farmers, particularly for judicious use of borrowed funds, it may to some extent prevent the farmer from falling into deep debt trap.

Social Support Mechanisms

In situation of economic distress, relief comes not only from financial support but also from social support. Social support system is an important instrument that averts seclusion and introversion. Deshpande and Gowda (Chapter 5) emphasize that social support provided to an individual by the family members, friends, relatives and village institutions, strengthen his belief in life, enhance entrepreneurial qualities and acts as a psychological insurance against any collapse of the person's faith in life. There is a general feeling that the social support has declined over the time due to disintegration of joint family system and deteriorating standards of the village institutions, particularly, the cooperative culture.

The resurrection of social support system is to a large extent outside the policy realm. But there are examples where State has played a decisive role in establishing and revalitalizing the social networks. The spread of SHGs is an example of social support system that as per the recommendations of the Radhakrishna Committee must be strengthened and federated so that a larger institutional network is available at the local level. For managing agrarian distress the Karnataka Agricultural Policy has considered the relevance of a welfare sub-system and has recommended creation of a farmers' welfare fund, a nodal department for the farmers and a wide social security system with facilities for the pension for the aged farmers. The Raithu Mithra Group (RMG) programme of Andhra Government has also been successful to some extent in re-building the mutual cooperation among the member farmers. Such kind of social groups and institutions need to be improved and extended for wider and effective social network

Institutional Reforms

The agriculture sector needs well-functioning markets to drive growth, employment, and economic prosperity in rural areas. Plethora of regulations on marketing of agricultural produce has created restrictive and monopolistic marketing structures, resulting in inefficient operation and high degree of marketing costs. The Guru Committee (June 2001) and the Inter Ministerial Task Force (May 2002) appointed by the Government, suggested widespread reforms in the agricultural marketing system. The reports have noted that the situation of control over agricultural markets by the State has to be eased to promote competitive agricultural markets and facilitate greater participation of the private sector.

In light of the recommendations, the scope of earlier State Agricultural Produce Marketing Committee (APMC) Act was broadened to include the objective of development of agricultural marketing in addition to its regulation under the Act. The new model Act, State Agricultural Produce Marketing (Development and Regulation) Act, 2003, has been redrafted to provide for development of efficient marketing system, a level competitive environment for private and cooperative sectors vis-à-vis regulated markets, promotion of agro-processing and agricultural exports and to lay down procedures and systems for putting in place an effective infrastructure for the marketing of agricultural produce. Appropriate legal framework is also provided to promote direct marketing and contract farming arrangements as alternative marketing mechanism.

Several State Governments have already amended their APMC Acts allowing varying degrees of flexibility but some are yet to follow (Box 3). Speedy implementation of the market reforms is essential to provide a conducive policy environment for marketing of agricultural produce. The Eleventh Plan Working Group on Agricultural Marketing has suggested a reorientation of policy paradigm involving shift in 'agricultural marketing' from the list of state subjects to the concurrent list for speeding up the progress of market reforms and evolving a unified national market.

The other area where institutional reforms are much needed is to modernize agrarian relations for equity and efficiency. Land is the prime resource of the vast majority of the poor in India deriving livelihood from agriculture. Given the existing stock of land resource, the issue of access to this resource is critical in determining the welfare of the farming community in India. An efficient and corruption free land administration, coupled with a dynamically adaptive land policy, has a vital role in increasing agriculture growth. Researchers have argued for liberalisation of lease market to activate the land market and improve the access of land to a typical Indian farmer who wants to overcome scale infirmities, especially in the interest of production

Box 14.3: Progress of Reforms in Agricultural Markets (APMC Act) as on March 2008

	Stage of Reforms	*Name of States/Union Territories (UTs)*
1.	States/UTs where reforms to APMC Act has been done for direct marketing, contract farming, and markets in private/co-operative sectors	Andhra Pradesh, Arunachal Pradesh, Assam, Chhattisgarh, Goa, Gujarat, Himachal Pradesh, Karnataka, Madhya Pradesh, Maharashtra, Nagaland, Orissa, Rajasthan, Sikkim and Tripura
2.	States/UTs where reforms to APMC Act have been done partially	(i) Direct Marketing: National Capital Territory of Delhi (ii) Contract Farming: Haryana, Punjab, and Chandigarh (iii) Markets in Private/Co-operative Sectors Punjab and Chandigarh
3.	States/Uts where there is no APMC Act and hence not requiring reforms	Kerala, Manipur, Andaman and Nicobar Islands, Dadra and Nagar Haveli, Daman and Diu, and Lakshadweep. Bihar has recently repealed the APMC Act w.e.f. 1.9.2006
4.	States/UTs where APMC Act already provides for the reform	Tamil Nadu
5.	State/UTs where administrative action is initiated for the reform	Mizoram, Meghalaya, J&K, Uttarakhand, s West Bengal, Pondicherry, Jharkhand, and Uttar Pradesh

Source: GOI, 2008

and trade efficiency (Sharma, Chapter 3). Hence, the existing tenancy laws should be amended to allow leasing-in and leasing–out land with provisions for security of both, the tenants and landowners. In the current context of agrarian distress, well functioning lease markets can contribute towards broad based rural development in several ways like, shifting land from less productive to more productive uses, encouraging rural households to take up non-farm jobs without the risk of losing their land and increasing investment incentives in that those who make such investment can enjoy the benefits even if they are not in a position to use the land personally.

Non-farm Employment

The multiple linkages between farm and non-farm sectors render equal importance to the growth of both the sectors for managing distress situation in agrarian economies. Wage employment programmes is the prime element of the popular. National Rural Employment Guarantee Act 2005 (NREGA). The Act provides for the enhancement of livelihood security of the households in rural India by ensuring at least one hundred days of guaranteed wage employment in every financial year to every household whose adult members volunteer to do public work-related unskilled manual work at the statutory minimum wage. Contrary to the envisaged potential of the policy for alleviating agrarian distress, it has come under scathing criticism by the World Bank in its 'World Development Report' 2009. The Report argues that NREGA acts as a policy barrier to internal mobility (rural urban migration), which is vital for lifting people out of poverty.

The efficacy of non-farm employment for growth of rural income depends on the ability of non-farm sector to absorb rural poor in high productivity income opportunities rather than low productivity residual activities. Self-employed and regular employment generating non-agricultural activities are important components of the non-farm sector, crucial in strengthening the base of rural livelihoods. The policy emphasis, therefore, should be on rural economic diversification to more value-added activities and non-agricultural activities. Agro-processing, particularly food processing, small and marginal enterprises (SMEs) and Khadi and Village Industries (KVI) offer increasing opportunities for the creation of productive employment in rural areas.

Several policy initiatives are needed to help promote food processing. The government should focus on R&D, information dissemination, extension and training and development of rural industrial estates (food parks). It should also create a policy environment, which supports creation and maintenance of required infrastructure by both, private and co-operative sectors. The government needs to encourage modern retailing to eliminate intermediaries and undertake a major export drive so as to create export opportunities for farmers and processors where price realisation is high and demands are huge. A major impediment to the development of a modern food processing industry is the present legal framework which is characterized by a multiplicity of laws and regulations related to food processing which are administered by different ministries and are often inconsistent with each other. These disparate laws should be integrated into a single Food Act, and a single Food Regulatory Authority should be set up to regulate all the health and safety aspects in the value chain of processed food. Central/State duties, taxes and levies on processed food, machinery, etc. need to be rationalized so as to reach international bench marks and make Indian food globally competitive.

The SMEs have a very significant role in rural economy in both, manufacturing and maintenance services. In addition they also play an important incubating role by developing entrepreneurship. The constraints to employment growth in SMEs are mainly in the form of inadequate finance and infrastructure availability and lack of skill and marketing that limit the scope of their business operation. Similar nature of problems also influence the performance of KVI. Suitable policy support is required to ease out the constraints faced by the SMEs and KVI so that these enterprises can serve as engines of productive non-farm employment growth in the rural areas.

Human Development

The policy measures focusing on improving input-output efficiency in agriculture, generating non-farm employment opportunities, etc. for addressing the agrarian problems are unlikely to yield desirable results unless due attention is paid to the development of human resources by way of empowerment through education, skill upgradation and better health.

The commitment of the state to disseminate knowledge among its citizens is reflected in Indian Constitution, wherein the Directive

Principle enjoins. However, nearly six decades after enforcement of the Constitution, rural literacy level could just touch half the mark and providing universal education to all remains a distant dream. The task for promoting elementary education and adult literacy is daunting. The existing education system is beset with multitude of problems like poor quality of infrastructure, high drop out rates, out-moded curriculum, untrained teachers and high rate of teacher absenteeism, especially in the rural areas. Empowering panchayats and citizens' education committees to oversee teachers' performance will help increase accountability.

As regards secondary education, intensive efforts are required to promote vocationalisation of secondary education, and incorporation of disciplines and areas which are more relevant in the present day scenario such that a disjunction does not take place between the educational system and the work place. The vocationalisation of not only secondary education, but also higher education is an important area calling for intervention.

Apart from education, health is an essential input for the development of human resources and the quality of life. Being a positive attribute of life, organization of health services to all people is considered the key step towards development. In India, after independence, there has been vast expansion of primary health institutions in the rural areas but the envisaged impact on rural health has not been made. The morbidity and mortality levels in the country are still unacceptably high. The public health care services are impeded by under-utilization of existing infrastructure due to inadequate number of medical and paramedical staff, unsatisfactory supply and maintenance of equipments, medicines and vaccines, and lack of commitment on part of the medical personnel. As a result of such inadequate public health facilities, a large segment of the population has to remain without a health cover that is within their economic means. To address infirmities and problems across rural primary health care, government has launched a seven year National Rural Health Mission (NRHM), which spans the duration of the 11th Plan. However, besides this initiative, there is a need to provide for health insurance and medi-claim services to the economically weaker agriculturist families. Also, greater NGO participation, contribution by other institutions of the civil society and creation of women groups need to be encouraged for the effectiveness of rural health.

Summing Up

Farm loan waiver, the most touted official measure to tide over agrarian distress is no more than a purely temporary relief that too with many exclusions and limitations. The solution to problems of farm households lies in a holistic approach that addresses the core issues in agriculture to make farming economically viable and absorbs the growing labour force in productive non-farm employment. A huge thrust on enhancing the TFP of agriculture, infrastructure development, risk management and institutional reforms is essential so that the farmers do not fall back into a debt trap, needing another loan waiver in the next few years.

References

RBI (2006), *Report of the Working Group to Suggest Measures to Assist Distressed Farmers*, Reserve Bank of India, Mumbai.

Mahendera Dev, S. (2008), Agriculture: Absence of a big push, *Economic and Political Weekly*, 12 April, 33-36.

Sainath (2008), 'Oh! What a Lovely Waiver', The Hindu, March 10.

Fan, S., P. Hazel and S. Thorat (1999), *Linkages between Government Spending, Growth and Poverty in Rural India*, IFPRI Research Report 110, Washington, DC.

Pal, Suresh and Derek Byerlee (2003), *The Funding and Organization of Agricultural Research in India: Evolution and Emerging Policy Issues*, Policy Paper 16, National Centre for Agricultural Economics and Policy Research, New Delhi.

Pal, S., A. Kumar and A. Singh (2000), Plan allocation and agricultural research intensity in India. Paper presented in *Workshop on Impact of Agricultural Research*, held at ICRISAT, Hyderabad, 20-21 February.

Acharya, S.S. (2006), Risk in agriculture: Some issues. *Agricultural Economics Research Review*, 19(1): 1-10.

GOI (2008), *Eleventh Five Year Plan, 2007-12*, Vol. III, Planning Commission, Government of India, New Delhi.

Mehta, M., K. Mukundan, R. Kesavan and S. Mukerjee (2003), *Sustainable Private Financing of Community Infrastructure in India*, Mimeo, World Bank.

Index